Smoke control in fire safety design

Smoke control in fire safety design

E. G. BUTCHER
B.Sc., M.Inst. P., F.I. Fire E.

and

A. C. PARNELL
F.R.I.B.A., F.S.I.A.D., F.I. Fire E., Dip. T.P.

Consultants in Fire Safety,
Fire Check (U.K.) Ltd

LONDON
E. & F. N. SPON

First published 1979
by E. &. F. N. Spon Ltd
11 New Fetter Lane, London EC4P 4EE

© *1979 Fire Check (U.K.) Ltd*

Printed in Great Britain by William Clowes & Sons Ltd, Beccles and London

ISBN 0 419 11190 5

Contents

Foreword

by KENNETH L. HOLLAND
CBE, QFSM, FIFireE

*H.M. Chief Inspector of Fire Services
for England and Wales
Home Office Fire Service Inspectorate*

Every year fire statistics demonstrate the tragedies caused by fires in the home, at work and at play. Most significantly they show the predominant cause of death to be the inhalation of smoke and toxic gases.

Designing for fire safety should today, therefore, be aimed at preventing smoke penetrating into escape routes so as to keep them clear, both for the escape of the occupants and for fire brigade access. Regulations have yet to be developed to define smoke movement in buildings, but there has been considerable research over the last twenty years to develop various systems for smoke control.

This book has brought together the results of such work and the experience of the Authors and others throughout the world, who have applied this knowledge in practical terms. Practising architects and engineers will find in these chapters encouragement to consider the challenge of designing for fire safety in buildings, so that people can safely escape in the event of a fire and so that the fire brigade may have a better chance to tackle the fire.

Acknowledgements

A great deal of the information set out in this book originated in publications of the Fire Research Station of the Building Research Establishment, and references to these publications are given, where appropriate, in the text. In particular, all the Nomograms in Chapter 3, Figs. 0.1 and 0.2 and Table 0.1 in the Preface, Fig. 1.4, and Fig. 4.12 are Crown Copyright and are published with permission of the Controller of Her Majesty's Stationery Office. Fig. 2.4 is printed by courtesy of the Honam Oil Refinery Co., Korea. Many of the details given in the case histories at the start of Chapters 2, 3 and 4 were supplied by the Fire Protection Association and the Authors gratefully acknowledge the assistance given by the Director of the FPA in this respect.

During the preparation of the manuscript, the Authors have had helpful discussions with many authorities both in the U.K. and overseas. These are too numerous to list here, but the help given is most sincerely appreciated and acknowledged. They would, however, like to acknowledge the helpful suggestions in respect of up-to-date information made by Mr. A. Heselden.

Finally, our thanks would not be complete without acknowledging the valuable help given to us by Mr. Gordon Harvey, who prepared all the illustrations, and Miss Valerie Jamieson who produced a complete text from all of the manuscript items sent to her from time to time during the preparation of this book.

Preface

Fire as a servant can be a good friend, but when it spreads uncontrolled in buildings it can be a vicious enemy. In the United Kingdom more than 1000 people died directly by fire in 1976 and more than £204 000 000 was claimed for direct losses in property and goods. The prime causes of death, however, have noticeably changed since the last War, and smoke and toxic gases are now causing more than 50% of the fatalities: 'The proportion of fires in dwellings in which furniture was ignited has remained constant over a twelve year period at just above 20%. Yet the statistics demonstrate that deaths from furniture ignited in the home have more than doubled and that the majority were caused by smoke and toxic gases.' [1]

Unfortunately, the smoke-producing agents in domestic furniture today are also processed, used or sold in every type of industrial, commercial and social building and, therefore, the risks from a small, but smoke laden fire are common to every enclosure. The Building Regulations and statutory control systems in many countries do not define, as a requirement, techniques for the design of buildings to ensure that smoke will not penetrate 'the protected routes' for escape in an emergency. It is therefore important to consider measures which will help to minimize this hazard, and an understanding of the mechanism of smoke production, and of the composition of the combustion products, is an important first stage. By visualizing the possible quality and density of smoke produced, often from quite small ignition sources, designers can appreciate the need for understanding the risks of rapid smoke movement within the buildings they design.

The following chapters, therefore, discuss the growth of a fire in the room of origin, and methods whereby it can be contained in that space and exhausted to atmosphere; as well as developing ideas for ensuring that such smoke will not penetrate into the escape routes. Case histories demonstrate recent and sometimes tragic experiences of ignorance of smoke movement

Table 0.1   *Causes of death in furniture fires in dwellings and in other fires classified by survey year*

Type of fire and causes of death	1962	1967	1970	1972
All fire deaths	667	779	839	1078
Burns	480	322	358	459
Smoke or toxic fumes	150	382	425	502
Other	37	75	56	117
Deaths in furniture fires in dwellings	156	212	270	289
Burns	90	59	47	79
Smoke or toxic fumes	56	140	213	189
Other	10	13	10	21
Other deaths	511	567	569	789
Burns	390	263	311	380
Smoke or toxic fumes	94	242	212	313
Other	27	62	46	96

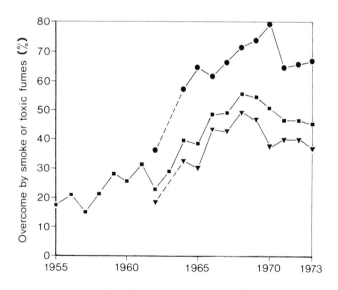

■ All fire deaths (buildings and outdoors)
● Deaths in furniture fires in dwellings
▼ Deaths in fires other than furniture fires in dwellings

Fig. 0.1   *Percentages of fatalities overcome by smoke or toxic fumes, 1955–1973.*

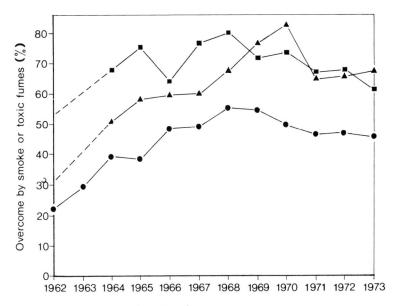

■ Upholstery fires (dwellings)
● All fire deaths (buildings and outdoors)
▲ Bedding fires (dwellings)

Fig. 0.2 *Percentages of fatalities overcome by smoke or toxic fumes, 1962–1973. (All fire deaths and fire deaths in dwellings due to ignition of bedding and upholstery.)*

and the dangers arising from the speed of travel. The design examples given in this book will provide guidance, and perhaps give more confidence, to architects and engineers to achieve more reasonable standards of safety in their building projects. However, the principles do not only apply to new building schemes and there will be many opportunities to apply the same concepts when upgrading our existing building stock, and even for maintaining the heritage of our historic buildings when improvements to their standards of safety are required.

REFERENCES

1. Chandler, S. E. (1976). Some trends in furniture fires in domestic premises. B.R.E. Current Paper 66/76, Fire Research Station, Borehamwood.

1 Smoke quantity and quality

1.1 GENERAL

The smoke produced at a fire will vary enormously, both from fire to fire, and from time to time in the same fire, and consequently in discussing the amount and nature of smoke produced it is only possible to speak in very broad terms.

The plume of hot gases above a fire will have many constituent parts, which will generally fall into three groups:

(*a*) hot vapours and gases given off by the burning material;

(*b*) unburned decomposition and condensation matter, (which may vary from light coloured to black and sooty);

(*c*) a quantity of air heated by the fire and entrained into the rising plume.

The cloud which surrounds most fires and is called *Smoke* consists of a well-mixed combination of these three groups and it will contain gases, vapours and dispersed solid particles.

The volume of smoke produced, its density and toxicity will all depend on the material which is burning and on the way it is burning; but each will probably depend on different factors in the overall behaviour and it is convenient to consider each separately.

In the following discussion concerning the *quantity of smoke*, the mixture of entrained air and smoky products of combustion will be treated as the total smoke produced. Its density and toxicity may depend on the material which is burning, i.e. the fuel, but the total quantity produced will depend on the size of the fire and the building in which it occurs. The nature of the fuel only affects the quantity of smoke produced in so far as the size of the fire depends on what is burning and the rate it is burning. This smoke may therefore be very dense or not so dense, but in any case it will be hot, and contain enough toxic products to be a danger to life whatever its density.

1.2 THE MECHANISM OF SMOKE PRODUCTION

The combustion of the solid materials in a fire involves the heating of those materials, usually by the adjacent burning material, and hot volatile combustible vapours are given off; these ignite so that above the fire there rises a column of flames and hot smoky gases, which because its density is lower than the cold surrounding air, will have a definite upward movement. As a result, the surrounding air is entrained into the rising stream and mixes with it (Fig. 1.1).

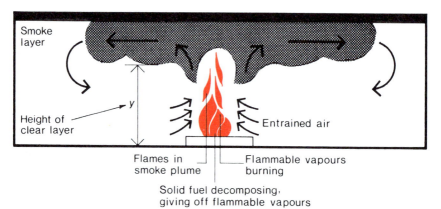

Fig. 1.1 *Production of smoke in a fire.*

Part of this entrained air will supply the oxygen needed for the combustion of the gases evolved by the decomposing fuel, and flames will be produced. However because the temperature in the plume is not high enough, and the mixing of the oxygen into it is not complete, the combustion of these gases will be incomplete and the dispersed solid particles which form the sooty component of the smoke will be produced.

At the height of the tips of the flames, the column of rising hot gases invariably contains much more air than is required or used for the combustion of the fuel gases, but by this time the excess air has been heated and well mixed with the hot smoky products of combustion, and so forms a large inseparable component of the smoke.

1.3 ESTIMATING THE VOLUME OF SMOKE PRODUCED BY A FIRE

Compared with the total volume of air entrained by the fire, the volume of the fuel gases is relatively small and it is therefore possible to say that *the rate of production of smoke by a fire is approximately the rate at which air is entrained (and contaminated) by the rising column of hot gases and flames.*

This rate of air entrainment will depend on:

(a) the perimeter of the fire;
(b) heat output of the fire;
(c) the effective height of the column of hot gases above the fire (i.e. the distance between the floor and the bottom of the layer of smoke and hot gases which form under a ceiling).

It has been shown [1] that the mass of gas entrained by a fire (and, therefore, the quantity of smoke produced) can be estimated using the following relation:

$$M = 0.096 P \rho_0 y^{3/2} \left(g \frac{T_0}{T} \right)^{1/2} \tag{1.1}$$

The meaning of the symbols used above are given in Table 1.1. The numerical values given in the third column of Table 1.1 are typical for a common fire situation when the flames in the smoke plume extend up to the layer of smoke under the ceiling, and it is appropriate to use these for the approximate calculation of the rate of smoke production unless other information is available for a particular circumstance.

Table 1.1 *Meanings of symbols and typical values of quantities used in Equation 1.1*

Symbol	Meaning	Typical numerical values which may be used to calculate smoke quantity
P	Perimeter of fire	As appropriate (expressed in metres)
y	Distance between floor and bottom of smoke layer under ceiling	As appropriate (expressed in metres)
ρ_0	Density of the ambient air	1.22 kg/m³ at 17°C
T_0	Absolute temperature of ambient air	290 K
T	Absolute temperature of flames in smoke plume	1100 K
g	Acceleration due to gravity	9.81 m/s²
M	Rate of production of smoke	in kg/s

Using the numerical values listed, the expression for estimating the rate of smoke production reduces to:

$$M = 0.188 \times P \times y^{3/2} \tag{1.2}$$

This reduced expression shows clearly that the rate of smoke production is directly proportional to the size of the fire (*P*) and dependent upon the height of clear space (*y*) above it.

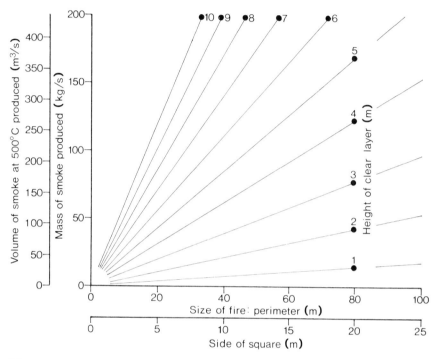

Fig. 1.2 *Smoke produced by fires.*

The graphs of Fig. 1.2 show the values to be expected for the rate of smoke production for values of fire size (*P*) and clear height above the fire (*y*) which might be expected if a fire occurred in a building.

The values for the size of the fire are shown as either:

(*a*) the perimeter of the fire (in metres); or,
(*b*) the length of the side of a square fire (in metres).

The values for the rate of smoke production are shown as kilograms per second, and the volume rate (in cubic metres per second) for smoke at a temperature of 500°C is also indicated.

The conversion of the mass rate of production of smoke to a volume rate can be made by calculation using the following information:

(a) the density of air at 17°C is 1.22 kg/m³
(b) the density of air (as smoke) at T°C is

$$1.22 \times \left(\frac{290}{T+273}\right) \text{kg/m}^3 \tag{1.3}$$

(c) the rate of smoke production in kg/s can be changed into m³/s by dividing by the density appropriate to the smoke temperature.

Equivalent values for mass rate of smoke production and volume rate in various units are given for a typical range of values in Table 1.2.

Table 1.2 *Conversion of mass rates of flow of smoke into volume rates of flow.*

Mass rate of flow		Volume rate of flow			
kg/s	*lb/s*	*m³/s*		*ft³/min*	
		at 20°C	*at 500°C*	*at 20°C*	*at 500°C*
200	440.8	163.9	436.9	346 928	925 354
100	220.4	81.9	218.5	173 464	462 783
90	198.4	73.8	196.6	156 308	416 399
80	176.6	65.6	174.8	138 941	370 226
70	154.3	57.4	152.9	121 573	323 842
60	132.2	49.2	131.1	104 205	277 670
50	110.2	41.0	109.2	86 838	231 286
40	88.2	32.8	87.4	69 470	185 113
30	66.1	24.6	65.4	52 103	138 517
20	44.1	16.4	43.7	34 735	92 557
10	22.0	8.2	21.8	17 346	46 278
9	19.8	7.4	19.7	15 631	41 640
8	17.7	6.6	17.5	13 894	37 023
7	15.4	5.7	15.3	12 157	32 384
6	13.2	4.9	13.1	10 420	27 767
5	11.0	4.1	10.9	8 684	23 128
4	8.8	3.3	8.7	6 947	18 511
3	6.6	2.5	6.5	5 210	13 852
2	4.4	1.6	4.4	3 473	9 256
1	2.2	0.8	2.2	1 735	4 628

1.4   LIMITING THE FIRE SIZE FOR THE PURPOSE OF DESIGN CALCULATIONS

In the general case of a developing fire in a building, the calculation of smoke quantities becomes almost impossible because the conditions are changing in a way which cannot usually be predicted. It has been shown that the rate of smoke production depends on the size (i.e. the perimeter) of the fire, so that as a fire grows and its boundaries spread, the rate of production will also increase.

On the other hand, the rate of smoke production also depends upon the height of clear space above the fire. As the fire develops, the smoke layer collecting beneath the ceiling will become thicker, the clear space above the fire will be reduced and therefore the rate of smoke production will become less and less. The magnitude of these two opposing effects will depend on the circumstances prevailing in the building at the time and there will, of course, be other factors operating to affect the amount of smoke produced as a fire develops, not least among these being the increase in heat output as the size of the fire increases. It is certain that even for a specified building it is not possible to quantify these variables and any attempt to calculate the rate of smoke produced or to design a smoke control system based on a general set of conditions is not realistic. However, one of the primary principles of fire protection is to make provisions in a building which will either limit the fire size or restrict its spread.

In a large building the installation of sprinklers will usually be specified, in which case it is possible to assume that the fire will be limited to a size which is approximately a 3 m × 3 m square. This assumption is justified by records which show that in a high proportion of fires in sprinklered premises the fire is controlled to a size approximating to the sprinkler head spacing. Consequently, in most smoke control designs sprinklers are specified as a firm requirement and the above fire size is assumed in any consideration of smoke control measures.

1.5   QUANTITY OF SMOKE PRODUCED

When the assumption is made that the fire size will be limited to a 3 m × 3 m square (or its equivalent circle 3.8 m in diameter), the rate of smoke production can be calculated from Equation 1.2 or found directly from Fig. 1.2, but Table 1.3 below gives typical values which have been calculated on the basis that the heat output from the fire is such that the flames will extend into the layer of smoke and hot gases which will form underneath the ceiling but above the fire.

Equation 1.2 and Fig. 1.2 give the smoke quantities in kg/s. The volume occupied by these hot gases depends on their temperature; close to the fire where the temperature may be as high as 500°C, 1kg of smoke will occupy about 2 m³; but a long way from the fire, where the smoke may have cooled to only slightly above ambient temperature, 1 kg will only occupy about 0.8 m³.

Table 1.3 *Rate of production of smoke from a 3 m × 3 m fire*

Height of clear layer (distance between floor and bottom of smoke layer) metres	Rate of smoke production		
	kg/s	Smoke volume at 500°C m³/s (ft³/min)	Smoke volume at 20°C m³/s (ft³/min)
2	6	13.1 (27 710)	5.0 (10 550)
2.5	9	19.6 (41 570)	7.5 (15 826)
3	12	26.2 (55 420)	10.0 (21 180)
4	18	39.2 (83 135)	14.9 (31 653)
5	25	54.5 (115 466)	20.7 (43 962)
6	33	71.9 (152 415)	27.4 (58 030)
8	51	111.2 (235 550)	42.3 (89 680)
10	71	154.8 (327 920)	58.9 (124 850)

In Table 1.3 the mass rates of flow have been converted to volume rates of flow for these two temperature conditions. It is clear from Table 1.3 that the volumes of smoke produced from quite a small fire are very large, and it is important to realize how quickly a building can fill with smoke.

When the level of the smoke layer which forms underneath the ceiling in a building extends downwards and reaches head level the occupants can be considered to be in extreme danger. This situation can develop very quickly, and in Table 1.4 the approximate times are given in which the smoke developed by a small fire (3 m × 3 m) will fill or partially fill buildings of various sizes. A glance at the figures in Table 1.4 shows that in all but the largest building the time to fill a room with smoke down to shoulder level is short, very short indeed in some cases.

Consider the smallest room in the Table, one whose floor area is 100 m². A room of this size and (say) 6 m high, which was set up as a small lecture room, could easily hold up to 100 people. If a fire occurred at the demonstrator's bench at the front (say, because of fractured glass apparatus causing a flammable liquid spillage, perhaps benzene, resulting in a flash fire), it would quickly spread and have a perimeter of 12 m in a very few seconds. Table 1.4 shows that such a fire in a lecture room of that size would fill it with smoke down to shoulder height (1.5 m) in 20 seconds. This will be before most people in the room have realized there is a fire, let alone started the escape movement. This is a startling illustration of the kind of dangerous situation which can very quickly develop because of the very large quantities of smoke which are developed by even a small fire (see Fig. 1.3).

Table 1.4   *Approximate times for a 3 m × 3 m fire to fill a building with smoke down to a given distance from the floor*

Building height	Building area 100 m² Distance of smoke from floor (m)			Building area 1000 m² Distance of smoke from floor (m)			Building area 10 000 m² Distance of smoke from floor (m)		
(m)	3	2	1.5	3	2	1.5	3	2	1.5
	(*Time in seconds*)			(*Time in minutes*)			(*Time in minutes*)		
4	4	11	17	0.7	1.8	2.8	6.9	18.4	28
5	7	14	20	1.2	2.3	3.3	11.5	23	33
6	9	16	22	1.5	2.6	3.6	15	26.5	36
8	12	19	25	2.0	3.1	4.1	20	31	41
10	14	21	27	2.3	3.5	4.4	23	35	44
15	17	24	30	2.8	4.0	4.9	28	40	49.5

The calculations made for Table 1.4 are only approximate, in that they ignore the time taken for the smoke to flow along under the ceiling until it reaches the boundary walls. This time is short: depending on the smoke temperature, the speed of the leading edge of the smoke layer may approach 1 m/s. Thus, compared with the time taken for the smoke layer to deepen it is reasonable to ignore it. Making this assumption, the time taken for a building of any size to fill with smoke may be calculated using the following relation derived by Hinkley. [1]

$$t = \frac{20A}{P \times g^{1/2}} \times \left(\frac{1}{y^{1/2}} - \frac{1}{h^{1/2}} \right) \tag{1.4}$$

t = time taken in seconds.
A = floor area of the building, room or compartment (m²).
P = perimeter of the fire (m).
y = the distance from the floor to the lower surface of the smoke layer (m).

h = the height of the building, room or compartment (m).
g = acceleration due to gravity (9.81 m/s²).

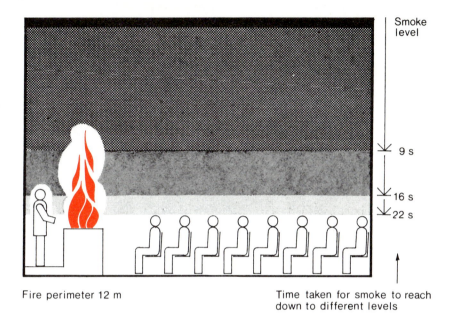

Fire perimeter 12 m

Time taken for smoke to reach down to different levels

Fig. 1.3 *Smoke levels in a small room. (Room height 6m ; floor area 100 m².)*

1.6 SMOKE QUALITY

The smoke produced by fires varies enormously in nature and content. It will vary in appearance from light-coloured to black and sooty, containing unburned decomposition and condensation products arising from the destructive combustion of the fuel. This variation is illustrated in the following examples. For convenience, the smoke density is expressed in terms of visibility, but the main features of this term are discussed in detail later in this Section. Given a small domestic lounge, of size 1250 ft³ (35 m³), then 1 lb (0.5 kg) of wood burning would produce enough smoke to reduce the visibility to about 3 ft (1 m), i.e. it would be impossible to see a hand at the end of an outstretched arm.

This is quite a small bundle of firewood!

And,   2½ oz (0.07 kg) of expanded polystyrene;
   or,   4 oz (0.1 kg) of foam rubber;
   or,   1 lb (0.5 kg) of polyurethane foam;
   or,   ½ pint (0.3 litre) of kerosene
are equivalent.

Any of these materials in the quantities given, if burnt in the same lounge, would reduce the visibility to less than 3 ft.

As another example, an ordinary polyurethane foam mattress with waterproofed nylon cover produced a large amount of smoke in a few minutes (Fig. 1.4).

Fig. 1.4 *Polyurethane foam mattress with waterproofed nylon cover – 5 min after ignition.*

A similar comparison can be given for commonly used building materials [2]. In Table 1.5, the result of burning a piece of building board 220 × 220 mm is shown, in a room of size 34 m³, for some 20–40 minutes. Again the results are expressed in terms of visibility. These examples serve to show that the variation in smoke produced is not only due to the use of different materials, but can also depend upon how the material burns.

Table 1.5 *Smoke produced by various building materials*

Material	Thickness (mm)	Visibility in metres when sample is burnt in 34 m³ room	
		Flaming combustion of sample	Smouldering combustion of sample
Plasterboard	9.5	17	15
Fibre insulating board	10.7	18	2.7
Chipboard	12.7	2.7	1.5
Birch plywood	6.4	4.2	2.3
Hardboard	3.7	4.2	2.2
Melamine faced hardboard	3.2	4	3.3
PVC faced hardboard	5.7	3	3.8
Rigid PVC	1.6	2.8	3
Polyurethane sandwich board	13	4.7	4
Glass fibre reinforced polyester (flame retardant)	3.3	1.5	1.6

Some of the constituents of the smoke will be gaseous and colourless (i.e. air, carbon monoxide, carbon dioxide, etc.); other constituents will be condensed vapours finely dispersed as droplets to form an opaque cloud (ranging from water vapour to tarry condensates); and the final type of component will be the dispersed solid matter (like soot), formed by the combustion of the fuel, which is too fine to remain as ash but which is dispersed into the energetically rising cloud of smoke.

In discussing the quality of the smoke there are two aspects to consider. These are: (1) the obscuration of light caused by the smoke and the consequent hazard of impaired visibility, and (2) the toxic nature of the constituent gases and vapours which present very real danger to life. These two aspects of the quality of smoke will be treated separately.

1.7 SMOKE DENSITY

Density is an important feature of smoke quality because it reduces visibility and so hinders the progress of a person escaping from a fire. If the person (or persons) concerned is in an unfamiliar place, reduced visibility can very quickly cause dangerous conditions.

The reduction of visibility depends on the composition and concentration of the smoke, the particle size and distribution, the nature of the illumination and the physical and mental state of the observer. Smoke density can be measured objectively by determining the reduction in the intensity of a light beam as it passes through a smoky atmosphere. This objective measurement can then be related subjectively to the reduction in visibility. The objective measurement of smoke density is usually expressed either in terms of the *light obscuration* or the *optical density* of the smoke.

1. *Light obscuration* is a measure of the attenuation of a light beam when it passes through an atmosphere of smoke. If I_0 is the intensity of an incident parallel beam of light and I_x is the intensity received by an optical measuring receiver (e.g. a photocell) after having passed through a path length (of smoke) x, then obscuration S_x, expressed as a percentage, is

$$S_x = 100\left(1 - \frac{I_x}{I_0}\right) \tag{1.5}$$

2. *Optical density.* The reduction of light as it passes through smoke will obey a logarithmic law. For instance, if in passing through 1 m of smoke the intensity of a parallel beam of light has fallen by 50%, then when this same beam of light passes through a second 1 m of the same smoke its intensity will have fallen by 50% of 50% i.e. to 25% of the incident light, and after passing through the third 1 m it will have fallen to 50% of 25%, to $12\frac{1}{2}$% of the incident light. This is known in optical work as Lambert's Law of Absorption. It can be expressed mathematically and is used to define the 'optical density of smoke' as the negative logarithm (to base 10) of the fraction of light which is transmitted through smoke of path length x, or:

$$OD_x = \log_{10}\frac{I_0}{I_x} \tag{1.6}$$

Thus the optical density of 1.0 means that 90% of the incident light has been obscured.

1.7.1   Relation between 'obscuration' and 'optical density'

The two different ways of expressing the objective measurement of smoke density may at times seem to be confusing, but they are related and percentage obscuration (S_x) can be converted to optical density (OD_x) by the relation:

$$OD_x = 2 - \log_{10}(100 - S_x) \tag{1.7}$$

provided the same path length in smoke is used for both the measurements of S_x and OD_x, and in this respect it should be noted that neither the light obscuration nor the optical density are absolute measurements of smoke density since they are both related to a path length of light in the smoke.

There is, however, a direct relationship between optical density and path length and to explain this we can quote Beers Law from analytical chemistry, which states:

If two solutions of the same coloured compound be made in the same solvent, one of which is, say, twice the concentration of the other, then the absorption due to a given thickness of the first solution should be equal to that of twice the thickness of the second, or:

$$l_1 c_1 = l_2 c_2 \qquad (1.8)$$

Where l is the thickness of the solution, c is the concentration of the solution and the subscripts 1 or 2 relate to the first or second solution.

Provided there is no change in the nature of the smoke particles, Rasbash [3] has suggested that Beers Law can be applied to smoke concentration, and there is a direct relationship between optical density (OD) and the product of the length of the light transmission path (x) and the concentration of the smoke (c),

i.e. $$OD_x = x \times c \times B \qquad (1.9)$$

where B is a constant depending on the nature of the smoke. It follows from this equation that optical density is directly proportional to path length (for the same sample of smoke) or:

$$OD_x = \frac{x}{y} \times OD_y \qquad (1.10)$$

where x and y are different optical path lengths.

The relationship between light obscuration and path length is not so simple but can be derived for specific cases by using the equations (1.7) and (1.10) above.

Although there is no formally agreed standard it is common practice to use a transmission path of 1 m in making objective smoke density measurements. This facilitates comparison between several sets of measurements.

It follows from Equation 1.9 that for a given smoke and a given path length, the optical density (OD) is directly proportional to the concentration of the smoke. This fact is important in considering all aspects of smoke density and in particular in considering smoke dilution. For instance:

If a smoke has an optical density of P for a 1 m path length, and if that same smoke is diluted with n times its own volume of fresh air (and well mixed), then the optical density per metre of the resulting diluted smoke will be P/n.

The ease with which optical density measurements can be transposed for different path lengths or for dilution of smoke is most useful, particularly

when, in the next section, the relation between optical density and visibility
is discussed.

1.7.2 Specific optical density

In recent work, Robertson [4, 5] has introduced the idea of specific optical
density and he suggests that this provides both a basis for estimating the
smoke production potential of different products and a means for estimating
the photometric density of the resulting smoke when expanded into different
room or building volumes. Methods for measurement of specific optical
density are described in an N.F P.A. Tentative Standard [6]. The property is
characteristic of the smoke production of a product for the sample thickness
used and under the thermal exposure conditions which are specified.

It is defined as:

$$D_s = D \times \frac{V}{AL} \qquad (1.11)$$

where D_s is the specific optical density at the time of measurement (D_m is
used to denote the maximum value of D_s during the test); D is the
measure of the degree of opacity, the negative logarithm (base 10) of
the transmission, ($D = OD_x$ of Equation 1.6); L is the length of the
light path over which the transmission is measured; V is the volume
of the chamber used to which the smoke is confined and measured; A
is the area of the specimen exposed to the specified heating conditions.

Equation 1.11 can be transposed to indicate either the maximum amount
of smoke produced by the material being considered:

$$aD_m = vD/L \qquad (1.12)$$

or, to show the Optical Density per unit path length of the smoke produced:

$$D/L = aD_m/v \qquad (1.13)$$

where a is the area involved in fire of the product being considered; L is the
optical path length; and v is the volume of the room in which the
smoke is produced *or* the volume occupied by the smoke.

1.7.3 Visibility

The subjective indication of smoke density is concerned with how far people
can see through smoke, and this may well be the most important feature
which decides the hazard presented by a given amount of smoke. Visibility
in smoke depends on many conditions, some of which are functions of the
smoke, others features of the environment, and others characteristics of the
observer. These conditions can be placed into three groups:

(a) Smoke: Colour of smoke; size of smoke particles; density of smoke; physiological effect of the smoke (i.e. its irritant nature).
(b) Environment: Size and colour of object being observed; illumination of object (intensity of light and whether back or front lighting).
(c) Observer: Physical and mental state of observer; whether in the controlled conditions of a laboratory investigation or whether in the panic or near-panic state of a real fire.

Information about the last group is the most difficult to obtain, and indeed it must be recognized that all the information generally available relates to the 'laboratory investigation'.

1.7.4 Measurements of visibility

The most recent and most comprehensive study of the relation between visibility and smoke density has been carried out by T. Jin [7], in which the effect of smoke quality and the illumination of the object were both considered. He found that the brightness of the illumination was a very important parameter and that the visibility measurements could be separated into two distinct groups: (1) those relating to forward illumination of the object; and (2) those relating to back illumination. His results for back illumination are the only ones known to the authors for this condition, but those for forward illumination agree reasonably well with the earlier work carried out by Malhotra [8] and by Rasbash [9].

The graph in Fig. 1.5 compares the visibility measurements made (for forward illumination) by the various workers when these are reduced to comparable values of optical density, and in view of the variables which must exist in this kind of work the agreement is remarkably good. Both Rasbash and Malhotra reported variations of up to 25% in the visibility recorded by the same observer for the same conditions but on different occasions. Jin reports variations of up to 30% depending on the several factors such as intensity of illumination and smoke colour. The difference in visibility between front lighting and rear lighting of the observed object is, according to Jin, about 250% or 2½ fold, rear lighting giving the greater visibility. In the graph of Fig. 1.5, a line represents the mean value of Jin's results, with a wide band showing the range of his variations.

Measurements for rear illumination are very few; Jin's work is probably the only systematic set of measurements available. His results for rear illumination are not shown on Fig. 1.5 but they can be deduced by applying the 2.5 factor indicated above.

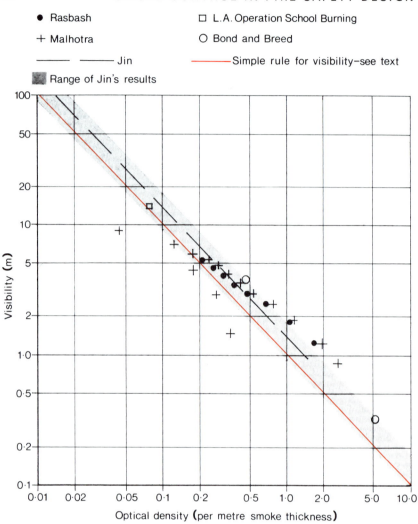

● Rasbash   □ L.A. Operation School Burning

+ Malhotra   ○ Bond and Breed

———— ———— Jin   ————Simple rule for visibility–see text

▨ Range of Jin's results

Fig. 1.5 *Comparison of visibility measurements.*

1.7.5 Simple rule for visibility

A glance at Fig. 1.5 will show that (for forward illumination results) it is possible to draw a single line which will pass centrally through the data plotted. This line can be used to give an approximate relationship between visibility and optical density of smoke which for most fire protection discussions or decisions will be adequate in its accuracy and which, because of its simplicity, is extremely valuable.

This approximate line indicates that in smoke whose optical density per metre is 1.0, the visibility is 1 m and for smoke with an optical density per metre of 0.1 the visibility is 10 m. It can be stated very simply in the following terms:

(*a*) For front illumination

$$\text{Visibility (in metres)} = \frac{1}{\text{optical density per metre}}$$

(*b*) For rear illumination

$$\text{Visibility (in metres)} = \frac{2.5}{\text{optical density per metre}}$$

These simple formulae are well worth remembering for all those who have reason to consider visibility in smoke conditions.

1.7.6   Practical values of optical density of smoke

The foregoing discussion of the relation between optical density and visibility may seem academic if the reader has no idea of the values of optical density likely to be experienced, and the following approximate values may be worth noting.

It is generally accepted that the dense undiluted smoke produced at a fire will have an optical density per metre of 10 or even greater. This, according to the simple rule stated above, will mean a visibility of some 10 cm (or 4 in) which is compatible with the remark 'unable to see a hand in front of one's face' but which would more truthfully be described as a 'nil' visibility. At the other end of the scale, it can be said that the minimum visibility acceptable on an escape route will be at least five metres (16 ft) which, again using the simple approximate relationship, will correspond to an optical density per metre of 0.2.

These are the two important levels of smoke density of which a building designer needs to be aware and, again according to the theory outlined above, in order to convert smoke whose optical density per metre is 10 to smoke of optical density per metre 0.2 it will be necessary to mix the original dense smoke with 50 times its own volume of fresh air.

1.8 SMOKE TOXICITY

All smoke from fires will contain gases which are toxic, and if exposure to smoke is prolonged its effect may be lethal. In some cases even short exposure can have fatal results. Rasbash [3] has reviewed the toxic hazards associated with smoke and his paper gives a very complete account of the present knowledge, particularly in regard to the dangers which may arise from burning plastic materials. Carbon monoxide is always found in smoke from fires and this gas has been identified in many post-mortem examinations (up to 40%) as being present in lethal amounts. Although it is only one of many potentially toxic constituents present in fire smoke, it is nearly always present in greater concentrations than others, and it is for this reason that most fatalities are due to (or ascribed to) its effect.

A recent report [10] from the Fire Research Station has indicated that in burning plywood (including examples treated with mon-ammonium phosphate fire retardant) the main toxic hazard is that due to carbon monoxide. This work is being extended to cover a selection of plastics materials commonly present in buildings. Similar investigations into the toxic problems created by burning plastics are being carried out at the University of Utah, USA, by Professor Einhorn [11] and his co-workers.

Several workers have identified other toxic components present in smoke from fires. Bunbury [12] has listed about fifty compounds given off by the destructive distillation of wood, and Woolley [13] has investigated the toxic products of combustion of plastics. The more common compounds found in fire smoke are listed in Table 1.6 [14] together with an indication of their

Table 1.6 *Toxic compounds which may be produced by the combustion of various materials*

Toxic gas or vapour	Source material
Carbon dioxide, Carbon monoxide	All combustible materials containing carbon.
Nitrogen oxides	Celluloid, polyurethanes.
Hydrogen cyanide	Wool, silk, leather, plastics containing nitrogen. Cellulosic materials. Cellulosic plastics, rayon.
Acrolein	Wood, paper.
Sulphur dioxide	Rubber, Thiokols.
Halogen acids (hydrochloric acid, hydrobromic acid, hydrofluoric acid, phosgene)	Polyvinyl chloride, fire retardant plastics, fluorinated plastics.
Ammonia	Melamine, nylon, urea formaldehyde resins.
Aldehydes	Phenol formaldehydes, wood, nylon. Polyester resins.
Benzene	Polystyrene.
Azo-bis-succino-nitrile	Foamed plastics.
Antimony compounds	Some fire retardant plastics.
Isocyanates	Polyurethane foams.

source materials. This Table is not by any means complete but it is sufficient to show the great variety of toxic compounds which may be present in smoke. The information available about the actual toxic effect of the various compounds released at fires is relatively scarce, and such knowledge as does exist is, necessarily, derived from experiments on animals.

Table 1.7 *Concentrations of various toxic gases which will be either dangerous for short exposure; or the maximum allowable for prolonged exposure*

Compound	Maximum allowable concentration for prolonged exposure (parts per million)	Concentration dangerous for short period exposure (parts per million)
Carbon dioxide	5000	100 000
Ammonia	100	4 000
Carbon monoxide	100	4 000
Benzene	25	12 000
Hydrogen sulphide	20	600
Hydrocyanic acid	10	300
Hydrochloric acid	5	1 500
Sulphur dioxide	5	500
Nitrogen dioxide	5	120
Hydrofluoric acid	3	100
Chlorine	1	50
Phosgene	1	25
Phosphorous trichloride	0.5	70
Acrolein	0.5	20
Bromine	0.1	50

In assessing the possible toxic effect of any substance in a fire, concern must be for the acute poisoning action due to exposure for short periods to high concentrations of the substance. Maximum acceptable concentrations of the toxic gas for prolonged exposure are more the concern of those interested in the hazards caused in industrial atmosphere. However, a very approximate relationship between these two concentrations can be deduced, as is illustrated in Table 1.7. The values for the two cases are only approximate. The compounds are listed in order of toxicity, the least toxic at the top of the Table. From the values given, it can be seen that the concentration regarded as 'dangerous for short exposure' is about twenty times the 'maximum allowable for prolonged exposure'. In no case is the ratio less than 1:20, and in some cases it is considerably more, which suggests that when the concentration which is 'dangerous for short exposure' is not known, taking a factor of twenty times the 'maximum allowable for prolonged exposure' will give a result which errs on the safe side.

It is clear from Table 1.7 that very small quantities of the toxic products of combustion are required to produce lethal conditions even for 'short exposure'. For instance, it has been stated [15] that exposure to an atmosphere containing 1% of carbon monoxide (CO) will give loss of consciousness in under 5 minutes and could, therefore, cause death in a very short time. Rasbash [3] has given values of CO concentrations measured in some experimental fires, and figures of up to 10% have been recorded.

To give some idea of the quantities of toxic gas that can be produced by the combustion of various materials, the masses and volumes of two such gases (carbon monoxide and hydrogen cyanide) produced by the burning of 1 kg of various materials, are given in Table 1.8. The figures given in the Table must only be regarded as approximate as the quantities of the two gases produced by combustion may depend on the particular conditions (e.g. ventilation) obtaining when the material was burnt. Since it is most likely that much more than 1kg of any of the fairly common substances given below will be involved in a fire, the volumes of toxic gas produced can be many cubic metres and so, even having regard for the large amounts of smoke and hot gases generated in a fire, concentrations of toxic gases of the order of those listed in Table 1.8 can be quickly exceeded.

Table 1.8 *Amounts of carbon monoxide and hydrogen cyanide given off by the combustion of 1 kg of various substances*

Substance	*Carbon monoxide*			*Hydrogen cyanide*		
	kg	*m^3 at 20°C*	*m^3 at 300°C*	*kg*	*m^3 at 20°C*	*m^3 at 300°C*
Cellulose (cotton)	0.50	0.4	0.8	–	–	–
Wool	0.23	0.18	0.37	0.12	0.1	0.19
Nylon	0.44	0.35	0.7	0.11	0.09	0.18
Acrylic fibres	0.3	0.24	0.47	0.26	0.21	0.42
Polyurethane foam	0.55	0.44	0.88	0.35	0.28	0.56

1.8.1 Synergistic effects of toxic gases

It is often suggested that the combined toxic effect of a mixture of dangerous compounds could be greater than merely additive, but until recently no information has been available on this point. In a recent publication Lynch [16] described some work on the problem, and he concludes that there is no synergistic effect when a mixture of carbon monoxide and hydrogen cyanide is inhaled.

1.8.2 Subjective nature of toxicity effects

This short section dealing with the toxicity of some of the products of combustion should not be concluded without the warning that the effects that exposure to toxic gases will have on any particular person may well depend to a considerable extent on the mental and physical condition of that person. For this reason, any statement about allowable or dangerous concentrations of toxic gases must be taken as indicating orders of magnitude rather than as precise figures. It could well be that in the mental stress conditions of an actual fire situation, toxic concentrations lower than those given above could have unfortunate if not fatal results.

REFERENCES

1. Hinkley, P. L. (1971). Some notes on the control of smoke in enclosed shopping centres. Fire Research Note No. 875, Fire Research Station, Borehamwood, England.
2. Thomas, P. H. (1969). The role of flammable linings in fire spread. *Board Manufacture,* **12**, 96–101.
3. Rasbash, D. J. (1967). Smoke and toxic products at fires. *Plastics Inst. Trans. J.,* Conf. Suppl. No. 2, 55–62.
4. Robertson, A. F. (1975). Estimating smoke production from rooms and furnishings. C.I.B. Symposium on the Control of Smoke Movement in Building Fires, Building Research Establishment, Garston, England.
5. Gross, D., Loftus, J. J. and Robertson, A. F. (1967). Method for measuring smoke from burning materials. S.T.P. No. 422, 166–204. American Society for Testing and Materials, Philadelphia.
6. N.F.P.A. Committee on Fire Tests (1974). Tentative standard test method for smoke generated by solid materials. N.F.P.A. 258–T, National Fire Protection Association, Boston.
7. Jin, T. (1970). Visibility through fire smoke. Building Research Institute, Tokyo, Report No. 30, March 1970 and Report No. 33, February 1971.
8. Malhotra, H. L. (1967). Movement of smoke on escape routes, instrumentation and effect of smoke on visibility. Fire Research Notes Nos. 651, 652 and 653. Fire Research Station, Borehamwood, England.
9. Rasbash, D. J. (1951). Efficiency of hand lamps in smoke. *Inst. Fire Engineers Quarterly,* **11**, 46.
10. Edginton, J. A. G. and Lynch, R. D. (1975). The acute inhalation toxicity of carbon monoxide from burning wood. Fire Research Note No. 1040, Fire Research Station, Borehamwood, England.
11. Birky, M. M., Einhorn, I. N. and Grunnett, M. L. (1973). Physiological and toxicological effects of the products of thermal decomposition from polymeric materials. Symposium: Fire Safety Research, N.B.S. Gaithersburg Md., U.S.A. Aug. 1973. U.S. National Bureau of Standards, Special publication. No. 411, pp. 105–24.

12. Bunbury, H. M. (1923). *The Destructive Distillation of Wood*. Benn Bros., London.
13. Woolley, W. D. and Raftery M. (1973) Toxic products of combustion of plastics. I.B.C.O. **11**, 2, 24–27.
14. Butcher, E. G. (1975). Plastics and fire. An increasing risk. *Fire Engineers' Journal,* **35**, 98, 24.
15. Henderson, G. Y. and Haggard, H. W. (1943). *Noxious Gases and the Principles of Respiration Influencing Their Action.* Reinhold Publishing Corporation, New York.
16. Lynch, R. D. (1975). On the non-existence of synergism between inhaled hydrogen cyanide and carbon monoxide. Fire Research Note No. 1035, Fire Research Station, Borehamwood, England.

2   Smoke movement and smoke control

2.1 INTRODUCTION

There are many case histories demonstrating rapid vertical smoke movement through buildings, and three have been chosen to demonstrate the need for an understanding of the fundamental principles.

The first – a fire at Orly Airport Terminal Building (1973) – shows how smoke and the ensuing flames find openings and weaknesses in the structural containment of the building to spread upwards. Once the smoke has entered the common circulation spaces of a building, especially such areas as escalator wells, open stairways, etc. which occupants frequently use for entry and exit, it will rapidly expand to the topmost available level and totally obscure all these familiar routes.

The second – No. 1 New York Plaza (1973) – is a case of rapid smoke and heat spread through concealed voids and ducts and its sudden and unexpected appearance ten floors above the floor of origin.

The third may seem to be a parallel story to the Orly disaster but in this hotel in Seoul, Korea (Christmas Day 1971) the smoke and heat rapidly spread to the very top of the building from which it could not gain release and, therefore, proceeded to layer downwards back through the building.

2.2 CASE HISTORY No. 1 – FIRE AT ORLY AIRPORT

Considerable disruption in airport and air traffic services for several days resulted from a fire on 3 December 1973 at Orly-South Airport (Paris). This report, kindly provided by the Centre National de Prevention et de Protection (the French FPA) shows how cavities, shafts and openings in a concrete floored building allowed fire, heat and smoke to spread from the basement to the top storey of a multi-storey building (Fig. 2.1)

2.2.1 Description of building

With a total floor area of 130 000 m², Orly-South airport building consists of six storeys above ground level and two below. The two basements are

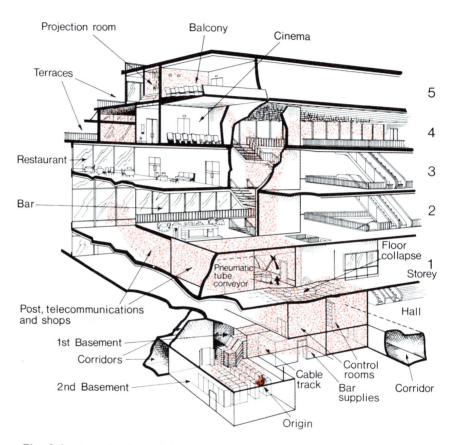

Fig. 2.1 *Spread of fire at Orly Airport.*

constructed of reinforced concrete. The concrete partition walls in the basement create separate compartments on both sides of a corridor. A group of staircases and escalators in the central part of the building link the various floors. The storeys themselves have relatively little compartmentation. The ground storey is a vast hall 200 m long.

2.2.2 Development of the fire

The fire started just before 15.00 h in the second basement in the low voltage sub-station. The fire spread to the cables in the service duct which led to a service point in the first basement. Unsealed shafts for cables allowed the fire to spread horizontally to a bar storeroom next to the service duct and vertically to control rooms in the ground floor storey.

On reaching the ground storey the fire destroyed all the equipment in two control rooms and caused serious damage to the metal framework of the floor above these areas. This resulted in distortion of the walls and floors, which allowed the gases and probably flames to pass directly up to the first storey.

The contents of the book shop, customs area and post office in the first storey helped to intensify the fire which then spread horizontally. Furthermore, a non-enclosed staircase ascending from the first to the fourth storey allowed the fire to spread rapidly to the higher storeys – due chiefly to a cavity about 10 cm wide between a timber lining and the wall which acted as a flue for the hot gases.

The central group of escalators allowed large quantities of hot gases to reach the fourth storey by another route. Above the escalators heat set fire to the plastic tiles of the false ceiling. The burning tiles falling from the ceiling increased the fire fighters' difficulties. Finally the fire reached the projection room of a cinema in the top (fifth) storey via an unenclosed staircase. The fire brigade arrived very quickly following the alarm, which seems to have been given by the fire detection system (fitted in places not permanently occupied). It was only at about 20.00 h and after five hours of fire-fighting in extremely difficult conditions – because of the poor visibility caused by the vast amount of smoke and the need to work with breathing apparatus – that the firemen managed to control the fire. The arrangement of the vertical shafts made it impossible for them to control the vertical spread of the fire. Stopping its horizontal spread in each of the storeys was made particularly difficult by the concealed spaces created by the false ceilings. Use of plastics in the cable coverings, the decorative elements, the furniture, etc. merely served to increase the difficulties of fighting the fire.

Damage caused directly by fire covered a surface area of about 6000 m^2; in addition 30 000 m^2 were damaged chiefly by smoke. An estimate puts the amount of damage at at least £3 million. Fortunately there were no casualties [6].

2.3 CASE HISTORY No. 2 – ONE NEW YORK PLAZA

Two lives were lost, 30 men were injured and almost ten million dollars damage was caused in the fire at One New York Plaza which occurred on 5 August 1970. The building was a 50-storey prestige office block, completed early in 1970, only partly tenanted and under a temporary certificate of occupancy. The fire occurred in the 32nd and 33rd storeys. It posed a number of questions about the design of lifts and their use in fires as well as showing the dangers of fire and smoke spread in a building thought to be of fire-resisting construction. This account is taken from a detailed report published by the New York Board of Fire Underwriters (NYBFU) with their kind permission.

2.3.1 Construction

The building has fifty storeys above ground and three basements. The first twenty storeys measure 286 ft × 222 ft (87 m × 67 m) and the tower above is approximately 286 ft × 143 ft (87 m × 44 m). A reinforced concrete centre core contains eleven lifts, five staircases, cloakrooms, air conditioning shafts and other utilities.

Walls are made up of aluminium panel window sections which also encase the outside columns. There is a six-inch concrete block curtain wall 28 in (0.7 m) high built on the outer edge of the floor slab. This wall is located in line with the centre of the wall columns so that the outer skin is 16 in (0.4 m) out from this wall. This separation creates vertical flues the height of the base or tower section which are interrupted at each floor level by an aluminium metal flashing designed to collect condensation and carry it through weep holes to the outside. The inside face of the curtain wall, the space between the windows and the space above the windows is insulated with a 1-in polystyrene foam board. This insulation is covered on the inside by gypsum board only where visible. There is no covering on it above the suspended ceiling. As a result, the protection between the concealed ceiling spaces of two floors consists of two one-inch thick pieces of foamed polystyrene and a thin sheet of aluminium. There are suspended ceilings of proprietary acoustic tiles and the ceilings voids which have no fire stops contain air ducts and electric cables and wiring.

2.3.2 Smoke and fire spread

The 32nd storey contained executive offices in the final stages of being prepared for occupation and the 33rd storey was also well furnished offices. The fire was first detected in the concealed ceiling space in the 32nd storey in the vicinity of, or directly under, a telephone control room on the 33rd storey.

The fire either started in this concealed space, where there were a substantial number of exposed cables, or in the offices nearby. Although smoke was noticed in the building at about 17.45 h the first call was not received by the New York Fire Department until 17.59 h. It was from a guard in an adjoining building. Most of the occupants of the building were either warned of the fire or saw smoke coming out of the air conditioning system. They were evacuated by the lifts or down the stairs.

The fire was drawn through the air conditioning system and probably spread to the polystyrene foam linings in the external walls, emerging as flaming droplets or gases into the office areas. The progress of the fire was accelerated by the foamed polyurethane upholstery of office furniture and the other combustible furnishings and contents of the offices. The flames were also drawn by the air conditioning system towards the centre core of the building, where there were lift shafts, staircases, air conditioning supply and return air shafts. A fire damper on the south wall of the core operated but another on the west wall was found partially open after the fire. The air supply fans were shut down at about 17.50 h by the operation of smoke detection equipment at the fans, which also sounded the alarm, but the extractor fans continued to run until 19.30 h. Because of this, smoke was drawn into return air shafts through the openings on the 32nd storey and through unsealed openings in the shaft. Positive pressure carried smoke to various storeys and it was sufficiently dense to require evacuation of the occupants.

The fire spread to the storey above through openings in the floor for electrical and telephone cables, around air conditioning ducts where they passed through the floor, and through the mail chute. As heat and flames passed through the openings around the air conditioning ducts, they melted and ignited polystyrene insulation in the outer wall. Fire Department appliances arrived at the scene within three minutes of the call but by this time the 32nd and 33rd storeys were raging infernos with so much smoke and heat that firemen could only operate in these areas for a short time. It was five hours before the fire was brought under control [7].

2.4 CASE HISTORY No. 3 – TAE YON KAK HOTEL, SEOUL, KOREA

The most disastrous fire for many years, in which 163 people died and 60 people were injured occurred at the Tae Yon Kak Hotel in Seoul, Korea. According to the report published by the National Fire Protection Association, the features which contributed to the tragedy included: the two interior staircases leading people only to the lobby on the first storey (where the fire broke out) and not to a place of safety; the stairways and shafts for utilities allowing smoke, toxic gases and fire to spread to all storeys and to cut off the only escape routes; openings in the division walls above the suspended ceilings permitting horizontal fire-spread between the offices and the hotel; and highly combustible interior finishes encouraging fire-spread throughout the building.

2.4.1 The building

The building with 21 storeys and a basement was only 18 months old. It was L-shaped, 160 ft (49 m) on the south side (the front of the building) and 140 ft (43 m) on the east side. The building was divided vertically by 8-in concrete block walls into two occupancies (Fig. 2.2). On the western side of the

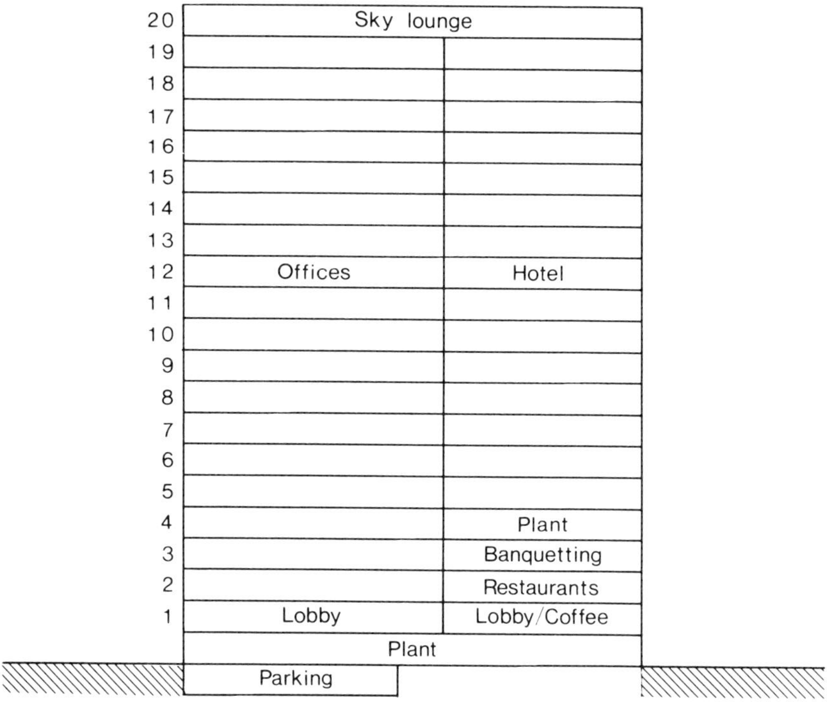

20	Sky lounge	
19		
18		
17		
16		
15		
14		
13		
12	Offices	Hotel
11		
10		
9		
8		
7		
6		
5		
4		Plant
3		Banquetting
2		Restaurants
1	Lobby	Lobby/Coffee
	Plant	
	Parking	

Fig. 2.2 *Vertical section through the Tae Yon Kak Hotel.*

building business firms occupied office space from the 2nd to 19th storey. The hotel part, in the east, contained 223 guest rooms and extended from the 5th to the 19th storey. The 'Sky Lounge' on the 20th storey extended over both office and hotel areas. There were two interior stairways, one in the hotel and one in the office area. From adjacent lobbies in the first storey there were two staircases, one serving all storeys in the hotel section and the other serving all but the top (20th) storey in the office section of the building. The two lobbies were separated by glass doors only. Both stairways were enclosed in plastered concrete block walls but the hotel stairs were open at the lobby level and in the three storeys directly above.

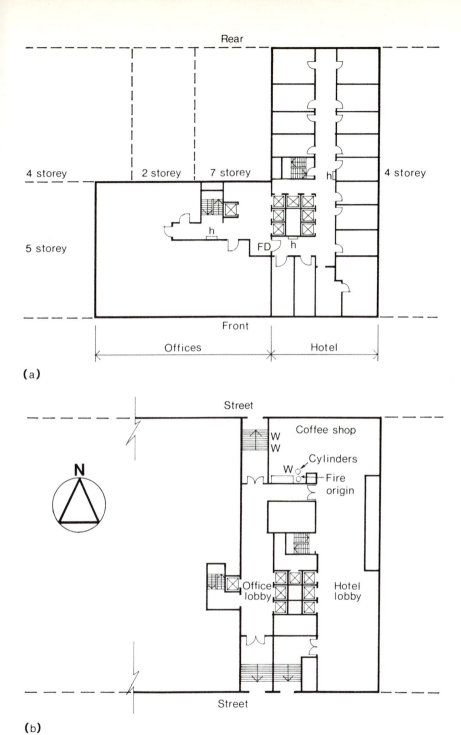

Fig. 2.3 *Floor plans of the Tae Yon Kak hotel; (a) typical floor plan of the 7th to 19th storeys; (b) the 1st storey. A 'W' marks the spot where each of three waitresses died.*

2.4.2   The start of the fire

There were approximately 200 hotel guests, 70 hotel employees and about 15 employees of the office firms in the building when the fire broke out just after 10.00 h. As it was Christmas morning many hotel guests were still in bed or just getting up. The fire started in the coffee shop adjacent to the hotel lobby (see Fig. 2.3) and involved liquefied petroleum gas, although the exact details of the cause of the fire are not known. Liquefied petroleum gas from a 20 kg cylinder supplied a two-burner stove on the counter. On the morning of the fire a spare 20 kg cylinder was beside the cylinder being used. It has been suggested that the spare cylinder could have failed first, releasing its contents which were quickly ignited or that a serious leak or opening in the relief valve caused the fire with the subsequent failure of the cylinder from flame contact. The body of the spare cylinder was later found ruptured and separated from its base; this cylinder had moved about 6 ft (2 m) and the force of the explosion had moved the counter outwards. Three waitresses died in the coffee shop. One was standing behind the counter near the cylinders; the other two, who were sitting facing the west wall, died in their seats. A hostess standing on the opposite side of the counter from the cylinders was seriously burned but she, and six other employees, escaped. There were no customers in the lobby or coffee shop at the time.

The fire engulfed the coffee shop immediately, spread throughout the lobby over the combustible interior finish and cut off escape down the hotel stairs. Smoke and toxic gases filled the building as the fire raced up the open stairway to the second and third storeys. On the second storey the ducts to vertical heating and air conditioning shafts were open and these spread smoke and heat through the offices and hotel.

Hotel guests were roused by the smell of smoke and by employees warning them of the fire. When they tried to leave they found corridors and stairways filling with smoke and heat.

2.4.3   The spread of the fire

The fire, which had started in the first storey, spread quickly up the next two storeys. Then it travelled up vertical shafts and ducting and early in the fire witnesses noted that the 'Sky Lounge' was involved. Then the fire progressed upwards from the third storey and downwards from the 'Sky Lounge'. The middle storeys did not begin to burn until the afternoon. Apparently, the building remained structurally sound following a total burn out in most areas (Fig. 2.4). Only one crack was noted in a column on the 17th storey although some superficial spalling of columns, beams and floor slabs was evident [8].

Fig. 2.4 *The Tae Yon Kak Hotel, after the fire.*

2.5 SMOKE MOVEMENT

There are two main factors which determine the movement of smoke and hot gases from a fire in a building. These are:

(*a*) The smoke's own mobility (or buoyancy) which is due to the fact that it usually consists of hot gases which are less dense than the surrounding air.

(*b*) The normal air movement inside the building, which may have nothing to do with the fire but which can carry smoke around a building in a positive way.

The relative magnitude of these two 'smoke moving' factors will depend on particular circumstances and will certainly differ from place to place inside a building. In general it may be expected that close to the fire factor (*a*) will dominate and as the distance from the fire increases (and the smoke gets cooler), so factor (*b*) will become more important.

The movement caused by the smoke's mobility is due to pressure differentials developed (*a*) by the expansion of the gases as they are heated by the fire, and (*b*) by the difference in density between the hot gases above the flames and the cooler air which surrounds the fire.

The normal air movement in the building can be caused by three separate factors:

(*a*) The stack effect – the pressure differential due to the air inside a building being at a different temperature from the air outside. This will cause the air inside the building to move upwards or downwards, depending on whether the air inside the building is warmer or colder than the air outside.

(*b*) The wind – all buildings are to a greater or lesser extent leaky, and wind penetration through these leaks contributes to internal air movement.

(*c*) Any mechanical air handling system inside the building.

All the features which contribute to smoke movement in a building are shown diagrammatically in Fig. 2.5.

2.6 PRESSURE DEVELOPED BY A FIRE IN A BUILDING

In spite of the apparently vigorous movement of gases above a fire the pressures developed are relatively small [1]. When a fire occurs in a building in which there are openings to atmosphere, the pressure near the floor will be slightly below atmospheric (i.e. air will be drawn into the fire) and near the ceiling the pressure will be slightly above atmospheric. Somewhere between these two positions there will be a level at which the pressure inside and outside the building is the same – this is called the *neutral plane*.

Fig. 2.6 shows diagrammatically the air flow in and the smoke flow out of a fire compartment, and indicates that for the door opening (R.H. side of room) air will flow into the room at the bottom of the door but smoke will flow out of the fire room at the top of the door.

An estimate of the height of the neutral plane is of considerable importance in smoke control. Its position depends mainly on the temperature of the gases and the dimensions of the openings into the fire room. Additionally its

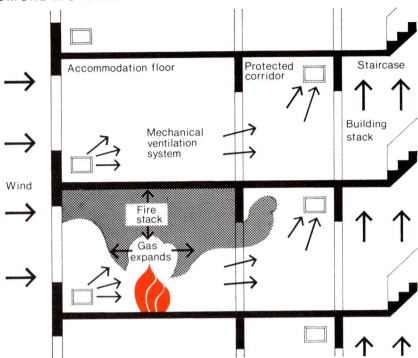

Fig. 2.5 *Factors affecting smoke movement.*

position may be variable according to whether the fire is growing rapidly or slowly, but for a reasonably steady fire it has been shown that the position of the neutral plane may be calculated using the following expression [2].

$$\frac{L_2}{L_1} = \frac{A_1^2}{A_2^2} \times \frac{T_F}{T_0} \tag{2.1}$$

Where L_1 = distance between the neutral plane and the bottom of the lowest opening.

        L_2 = distance between the neutral plane and the top of the highest opening.

        A_1 = Area of openings below the neutral plane.

        A_2 = Area of opening above the neutral plane.

        T_F = Temperature in the fire chamber in K.

        T_0 = Temperature outside the fire chamber in K.

   Note: $K = °C + 273.$

The magnitude of pressure differential developed over a fire will depend on the length of the column of hot gases above it. (Any pressure developed by

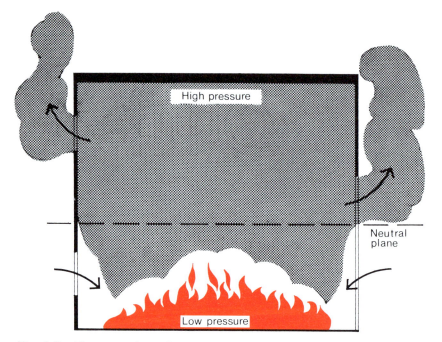

Fig. 2.6   *Flow into and out of a compartment full of hot gases.*

the expansion of the gases due to heating by the fire will be rapidly released because the fire room will not be a sealed volume.) Thus, so long as air appears to be entering the building the pressure difference between a point above the fire and atmospheric pressure remote from the fire will be the difference in 'heads' of the hot and cold air.

The value of the pressure differential so developed can be estimated by using the graph shown in Fig. 2.7, which relates height above neutral plane and temperature of the column of hot gas above the fire to the pressure developed [3]. It must be realized that at best it is only possible to make an approximate estimate of the pressure so developed, because an exact knowledge of the temperature of the hot gases is rarely available, and indeed in practice there will be a temperature gradient in the rising plume of smoke and hot gases above a fire.

It will be seen from Fig. 2.7 that the pressure developed by a fire in a building is in general very small, and even with a very tall compartment it will only be of the order of 100 Pa – which is 1/1000 of atmospheric pressure. As another example, at the height of the top of a door (say, 2 m from the floor) when the neutral plane is 1 m from the floor, the pressure due to a fully developed fire will only be 5 Pa.

Experimental measurements of the pressures developed in fires in buildings

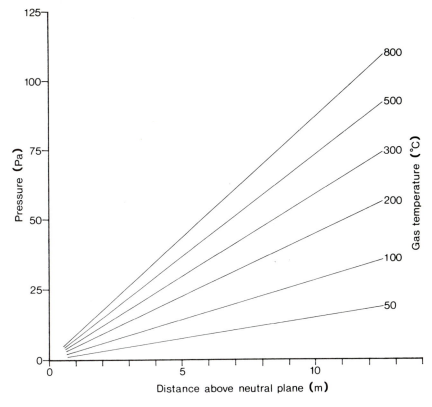

Fig. 2.7 *Pressure developed above a fire.*

have been made by several workers [4, 5] and the measured values agree reasonably well with the values suggested in Fig. 2.7. The pressure differentials which are caused inside a building by the natural environmental conditions, or by the mechanical ventilation system, can be comparable in magnitude with those developed in a fire and if this fact is realized and design features are incorporated into a building, then the movement of smoke and hot gases from a fire in that building can be controlled in such a way that fire and smoke spread will be minimized, and the safety of the occupants in making their escape ensured.

2.7 SMOKE CONTROL

It has been shown in Chapter 1 that very large quantities of dense toxic smoke will always be produced in a fire in a building and it is now suggested that the building designer must recognize that this will constitute a major hazard to the occupants. Consequently his design must incorporate features which will minimize the dangers.

In planning a building, there are two parts to the path along which an occupant must move to reach safety in the event of a fire. These are (1) the movement across the compartment in which the fire is occurring and (2) the subsequent movement along a path which is protected from the fire area by structural separation. The smoke control possibilities in the two cases are very different.

In the first, i.e. in the immediate vicinity of a fire, there must inevitably be a very large quantity of very dense smoke, and control measures can only be aimed at keeping that smoke confined laterally and at a high level so that occupants can move to safety in the clear space beneath it. There will almost certainly be a limit imposed on the distance over which the building occupant will have to travel in this situation.

In the second, i.e. in those parts of the building which are structurally separated from the fire area (such as corridors, lobbies and staircases) there is generally no restriction on travel distance, and movement by the occupants through these spaces must be possible for a reasonably long period (that is, during all stages of the fire). For this reason a smoke control system for these spaces must ensure that they are kept completely clear of smoke, or that any encroachment of smoke is so slight as to present no visibility or toxicity problem.

In very large, complicated buildings or a building complex it may be impossible to divide the escape route into the two clear cut stages described above, and it may well be found that parts of the escape route which must be kept usable for large numbers of people for reasonably long periods of time cannot, by reason of the functional purpose of the building, have full structural separation from the effects of a fire. An example of such a situation is the pedestrian mall in a covered shopping centre. The smoke control measures required here need special consideration.

In view of the different requirements and conditions imposed by various parts of the building, the description of smoke control methods and principles will be dealt with under three headings:

1. Smoke control in the actual fire area (e.g. by roof venting, or smoke extraction), (Chapter 3).

2. Smoke control on an escape route which does not have complete structural separation from the fire area (e.g. the pedestrian malls in covered shopping centres), (Chapter 4).

3. Smoke control on protected escape routes (e.g. by pressurization), (Chapter 5).

2.7.1 *Smoke control and fire size*

In any fire, the quantity and rate of production of smoke will depend enormously on the size of the fire, and it may be necessary in the design of a smoke control system to assume a likely fire size. Using the three headings given above, the impact of *fire size* on the smoke control design is as follows.

1. *In the actual fire area.* The design of the system and the sizing of the vent areas or the extraction rate require the assumption of a maximum fire size. This gives rise to three possible situations:

(*a*) The size of a possible fire must be restricted by the installation of a sprinkler system.

(*b*) The combustibles in the likely fire area must be separated into discrete sections, each of a limited size, and with adequately wide gangway separation between each section so that fire spread can be considered as being limited.

(*c*) If neither of the above conditions are possible, and if it must be accepted that the fire could grow to fill the whole building, then it must also be accepted that the smoke control system will only be effective for a short period during the early stages of the fire, and it must be estimated that this short period will be long enough to travel across the fire floor to a place of safety.

2. *In an escape route with incomplete structural separation.* A pedestrian mall in a covered shopping complex must remain usable for a long period of time, and since the smoke control system can only be designed for a given fire size, the design of the fire area must incorporate features which will limit the fire size. This means that in such a complex all of the shop areas must be sprinklered. The suggestion that the fire size could be limited by spatial separation of the combustibles is not practicable for the functional requirements of shops and the time limitation imposed by a growing fire is not acceptable in this type of building.

3. *On protected escape routes.* The smoke control system is designed to completely prevent smoke from entering the protected escape route by raising the pressure in those routes. The design criteria adopted are based on the maximum possible size of fire and, therefore, in this case no assumptions are made about fire size in the smoke control design calculations.

REFERENCES

1. Hobson, P. J. and Stewart, L. J. (1972). Pressurization of escape routes in buildings. Fire Research Note No. 958, Fire Research Station, Borehamwood, England.
2. McGuire, J. H. (1967). Smoke movement in buildings. *Fire Technology,* **3**, (3), 163–174.
3. Hinkley, P. L. (1971). Some notes on the control of smoke in enclosed shopping centres. Fire Research Note No. 875. Fire Research Station, Borehamwood, England.

4. National Fire Protection Association (1959). *Operation school burning*. Official report on test conducted by the Los Angeles Fire Department. U.S.A.
5. Butcher, E. G., Fardell, P. J. and Clarke, J. (1969). Pressurization as a means of controlling the movement of smoke and toxic gases on escape routes. Paper 5, Fire Research Station Symposium No. 4, *Movement of smoke in escape routes in buildings*. Watford 1969. HMSO. 1971.
6. *Fire Protection* No. 106, December 1974. Fire Protection Association.
7. *Fire Protection* No. 97, January 1973. Fire Protection Association.
8. *Fire Protection* No. 96, October 1972. Fire Protection Association.

3   Smoke control in the fire area

3.1 INTRODUCTION

As buildings become larger and their use more complex, the requirement for large undivided production or trading areas becomes more frequent, and in some circumstances paramount to the successful use of the building. In these cases the use of fire resisting dividing walls, the traditional method of limiting fire and smoke spread, becomes unacceptable. For such buildings the high level removal of smoke and hot gases is a method of preventing the lateral spread of fire and smoke and it may well be that in some circumstances this is more effective than dividing walls. It is particularly valuable when a combustible ceiling is present.

In single storey buildings, roof vents can be used to allow the smoke and hot gases to escape, but the principles can also be extended to cover the use of mechanical extraction systems in multi-storey buildings, and for information on the required size of the extraction units for this purpose, reference should be made to Chapter 4.

When the contents of a building are so combustible and so distributed that fire would spread rapidly throughout the entire building before any extinguishing action was possible, venting is of little use in reducing fire spread. On the other hand, venting is particularly valuable when the combustible contents are disposed in such a way that fire spread at low level is mainly caused by heat radiated downwards from smoke and hot gases below the roof. An extensive and comprehensive investigation into the principles, use and design of roof venting systems has been carried out by Hinkley and Thomas [1, 2, 3] and many of the details given in this Chapter have been taken from their work.

3.2 FACTORS IN THE DESIGN OF ROOF VENTING SYSTEMS

The quantity of hot gases produced by a fire and the way in which they move round a building will depend on several factors, such as the size of the fire, the size, shape and height of the building and the sizes and positions of openings into and out of the building. However, most of these factors are functions of the building concerned and for any given design they can be considered as known and generally fixed in magnitude. The one unknown is the size of the fire. In their discussion, Hinkley and Thomas have considered three sizes of fire and they suggest that if a growing fire is considered as being composed of a series of steady state conditions, then by using one or more of the three fire sizes they have chosen, together with the appropriate method of calculation associated with each, sufficient information can be obtained to cover most situations.

3.2.1 Fire development

In the early stages of a fire the rate of production of smoke and hot gases is such that all of it can be removed through vents suitably placed in the roof (Fig. 3.1). As the fire grows in size, the rate of gas and smoke production increases and a stage will be reached at which all the gases cannot be removed through vents in the roof because the vent area required would be too large to be considered practicable. Consequently, a layer of hot gases will be formed below the ceiling. This layer will spread sideways over a

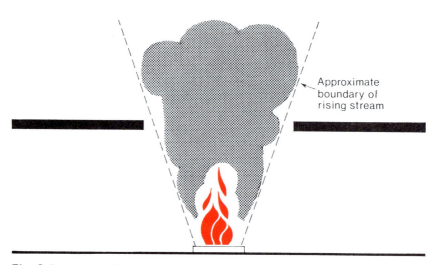

Fig. 3.1 *Stream of hot gases passing through a large vent.*

considerable area unless it is contained, and its depth will increase with the further progress of the fire.

This increasing depth of the smoke layer will have two effects: (1) a greater pressure will become available for driving the gases out through the vents; and (2) the free height above the fire will get less, thus reducing the rate of increase of air entrainment (see Chapter 1) which will, in turn, reduce the rate of development of the fire. The design procedure for a roof venting system endeavours to ensure that the smoke produced by the fire and flowing into the smoke layer will balance the smoke escaping from the vents with no increase in the depth of the smoke layer, the maximum depth of which will have been predetermined by the design and construction of smoke reservoirs beneath the ceiling.

3.2.2 Deciding the appropriate fire size

It is one of the major difficulties associated with the design of a roof venting (or smoke extraction) system that a given fire size has to be assumed in the design calculations. The most satisfactory way of dealing with this problem is to install a sprinkler system. The record of performance of such an installation is such that it can be assumed with reasonable certainty that the fire will be limited in size, the exact size depending on the sprinkler head spacing; but in most smoke control calculations the limiting size of a fire in a sprinklered building is assumed to be 3 m × 3 m. However, it may be argued that the roof venting (or smoke extraction) system itself should be capable of controlling the spread of the fire, at least until effective extinguishing action can be established, without incurring the additional expense of a sprinkler system.

If this idea is to be supported, it must be understood that the major cause of fire spread across the floor of the building is heat radiated downwards from the layer of hot gases beneath the ceiling. Roof venting will limit fire spread because it limits the spread of hot gases under the roof. On the other hand, if the major cause of fire spread is due to flame progressing sideways at floor level through readily combustible material, roof venting will not prevent this, it will only slow up the sideways movement because it will limit the extent of the heat radiated downward which will be only one factor in the sideways development of the fire.

The decision as to the appropriate fire size for design purposes may be made, on the following considerations:

(*a*) If a sprinkler system is installed the fire can be assumed to be limited to a 3 m × 3 m size (unless the sprinkler head spacing is very different from this).

(*b*) If no sprinklers are to be installed, but the combustible material on the shop or factory floor is divided into discrete sections with spaces in between, then the smallest fire size will be not less than the largest isolated pile of combustible material (such as a stack of packing cases, a pile of material or stores to be used, a single machine or a tank of flammable liquid).

(*c*) When the boundaries of the fire cannot be defined as in (b) above, then the size the fire will reach before effective extinguishing action is established must be estimated. This may be very difficult; it will depend on the rate of spread of the fire, on the fire size when it was detected and on the attendance time of the fire brigade. Two of these factors may be reasonably well known. If automatic fire detectors are installed the size of the fire when it is detected can be estimated and the attendance time of the fire brigade is usually known approximately. But the rate of fire spread will vary greatly with the total fire load, with the nature of the combustible materials and with the way in which it is distributed in the building.

(*d*) If it is difficult to assess the size of the fire after considering the techniques described in (c) above, an alternative approach may be possible. Decide on the minimum acceptable height for a clear air layer at ground level and estimate the size of a fire which can be considered in this building for the proposed roof venting system. A judgement can then be made on the advantages of providing such a venting system. In this context the minimum acceptable height should be taken as 2 m or the height to the top of any opening leading to an adjoining area.

3.2.3   Designing for a small fire

In this context a small fire is defined as one which:

    (*a*) has an area which is only a small part of the floor area of the building;
    (*b*) has flames which are short, compared with the height of the building.

This applies to the early stages of almost every fire up to the point when it should be detected, and any automatic action as a result of that detection initiated (such as vents opening, sprinklers operating, alarm raised, etc.).

The hot combustion products from the fire will rise and cold air will be entrained into this upward stream. Since the flames are short, no further heat will be produced in the rising stream and the temperature of the gases will fall because of the entrained air, whose mass will, in general, be many times greater than the mass of the gases emitted by the fire. These rising gases will reach the ceiling, and in the absence of vents will spread out to form a layer under the ceiling which rests on top of the cold clear air below and which mixes quite slowly with it; however, this layer will get deeper and deeper until the whole building is filled.

Since the fire is small it is possible to prevent the smoke filling the building by opening a vent in the roof (directly above the fire) of such a size that all, or nearly all, of the hot gases will flow directly out through this vent (Fig. 3.1). Data published by Yih [3] estimates that if the fire on the ground is very small compared with the height of the building, about 90 per cent of the hot gases will pass straight through a vent having a diameter equal to one third of the height of the building, irrespective of the rate of heat generation. Thus, in a 6 m (20 ft) high building a small fire, however severe, will require a vent directly over the fire which is larger than 2 m × 2 m if most of the hot gases are to go straight out, and with a larger fire, even if it is burning less rapidly, larger vents will be required.

It is clear that this is not a practical possibility for two reasons:

(*a*) the vents are too large to be economic;
(*b*) the position of the fire is not known so the vents cannot always be immediately over the fire.

In a system design, therefore, it is accepted that a layer of smoke and hot gases will form beneath the ceiling but by doing so the discharge of smoke through the vent is increased, since the layer develops a pressure head to drive smoke out, and at the same time the rate of production of smoke will decrease because the clear height above the fire is becoming smaller (Fig. 3.2). At some stage an equilibrium will be reached when the smoke venting is equal to the smoke production, and this is the basis for the design of a roof venting system. This assumes that adequate cold air can be drawn into the building at low level to replace the hot gases flowing out, and unless the openings for this purpose are sufficiently large the maximum venting capacity of the roof openings will not be realized.

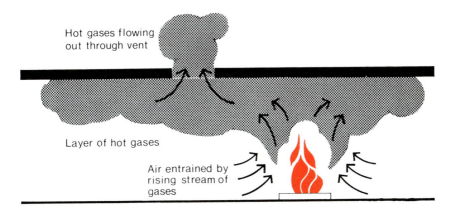

Hot gases flowing out through vent

Layer of hot gases

Air entrained by rising stream of gases

Fig. 3.2 *Formation of a layer of hot gases.*

Nomograms 1 and 2 (taken from [2]) enable calculations to be made of the areas of vent necessary to exhaust the hot gases produced by a small fire (defined here as a fire whose mean diameter is less than half the height of the layer of clear air above the floor). To use Nomogram 1, the fire is envisaged as a point source of heat, but for actual fires, which are generally too large to be regarded as point sources of heat, an equivalent point source can be defined at a distance below the fire which depends on the size of the fire (Fig. 3.3). The distance below the floor for this notional point source is given by:

$$d = 1.5\sqrt{A_f}$$

where d = notional distance below the floor of the point source
         A_f = area of the fire.

In using the Nomogram the building height is replaced by the distance measured from this notional point source.

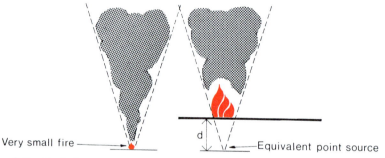

Fig. 3.3 *Equivalent point source of a small fire – some distance above the fire the two streams of hot gases behave in the same way.*

3.2.4 Designing for a large, but not fully developed fire

As the small fire of the previous section grows, its area increases and the flames get longer, probably reaching (or nearly reaching) the ceiling. It is no longer possible to regard it as a point source of heat. The description of the formation of a smoke layer under the ceiling still applies; but the theory of the behaviour of the smoke is now based on the calculation of the entrainment of air into the flames. The calculations so developed apply to all but the largest of fires, and Nomogram 3 of this chapter enables vent areas to be calculated.

There is, however, an upper limit to the absolute size of fire to which it applies. This arises when the vertical area* of the flame boundary over which

---

* This is the area of the low level openings to a building which is completely involved in fire; otherwise it is the height of the clear layer multiplied by the perimeter of the fire.

air entrainment takes place becomes small compared with the area of the fire. This section applies to fires which do not fill the whole building, compartment or division, but may well cover the whole of the floor area below a roof section (if this is divided as later described).

3.2.5   Designing for a fully developed fire

When a building or division of a building is completely involved in fire, venting cannot be used to maintain a clear layer of air which will enable personnel to enter for fire fighting or rescue; but it can be used to ensure that air flows inwards through all openings in the walls while flames and hot gases flow out only through the vents. Yokoi [4] has developed the theory which is applicable to this fire size.

In a building completely involved in fire the pressure distribution will be on the lines shown in the idealized representation of Fig. 3.4. The pressure near the ground is below atmospheric, so air will flow in, and near the ceiling it is above atmospheric, so smoke and flame will be driven out. Somewhere between these two levels the pressure in the fire room will be equal to atmospheric so at that level there will be no inflow or outflow. (See Fig. 3.4.) Opening vents in the roof will raise the neutral plane, and making openings at a lower level will lower it; the design idea is to ensure, by correct sizing of vents and low level openings, that no flames or hot gases will flow into an adjacent part of the building through openings in an otherwise fire resisting wall which has been used to subdivide the building into smaller divisions.

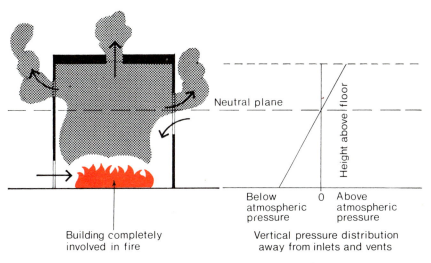

Fig. 3.4   *Idealized pressure distribution in a building with small openings.*

Nomogram 4 (a) and (b) allows the areas of vents for this purpose to be calculated, and it is important to consider all low-level openings (including leaks in the building fabric) which lead into the vented division, not only those which connect it with an adjacent division.

3.3 THE NOMOGRAMS

3.3.1 Using the nomograms

In using the nomograms the following definitions will be found useful:

Centre of vent. The geometric centre of the upper surface of the vent. Special types of ventilators, e.g. ridge or aerodynamics ventilators may have a centre at some other point. Information must be sought from the supplier.

Bottom of layer of hot gases. The position of the lower face of the hot gas layer assuming it to be undisturbed and uniform in temperature. In many cases this lower face may be irregular instead of flat in which case an average position may be used without introducing any appreciable error.

Effective height of the ceiling (h_c). The distance between the floor and the centre of the vents.

Effective depth of the layer of hot gases (d_b). The distance between the centres of the vents and the effective bottom of the smoke layer.

This distance will usually be either:

(*a*) equal to the effective depth of roof screens used to limit the lateral spread of this layer; or,

(*b*) limited by the required depth of clear air near the floor.

Effective depth of a roof screen. The distance between the centres of the vents and the lower edge of the screen.

Coefficient of discharge (C_v). The factor by which the actual rate of discharge through an orifice such as a vent differs from the rate of discharge calculated from energy considerations.

Measured free area of a vent (A_v)*. The area of free opening of the vent.

Aerodynamic free area of a vent ($A_v C_v$).* The product of the measured free area and the coefficient of discharge.

---

* *Note*: The nomograms have scales for both the measured free area and the aerodynamic free area (assuming a coefficient of discharge of 0.6). The aerodynamic free area is sometimes quoted by manufacturers.

Effective point source of a small fire. Small fires which are approximately square or circular in shape are regarded as originating at an effective point source a distance (*d*) below the base of the fire, where:

$$d = 1.5\sqrt{A_f}$$

Where A_f is the area of the fire. This may overestimate the value of *d* for a real fire and so tend to give too large a vent area. However, the error will be small.

Perimeter of the fire (W_f). The distance measured round the base of the fire provided it is roughly square or circular.

The fire conditions appropriate to each nomogram have already been discussed but for convenience are summarized in Table 3.1.

Table 3.1 *Conditions for which various Nomograms may be used*

Fire size	Definition of fire size	Conditions for cold air supply	Nomogram number
Small fire	Mean diameter of fire less than half the height of the clear layer.	Entrainment area† greater than half area of the fire.	1 2
Large but not fully developed fire	Mean diameter of fire greater than height of clear layer.	Entrainment area† greater than half the area of the fire.	3
Fully developed fire	Completely fills building or division of building – no clear layer.	Total area of low level openings less than 1/20 of the area of the fire.	4

† The Entrainment area of the fire is (for fires which do not completely fill the building or division of the building) the product of the perimeter of the fire and the height of the clear layer.

Nomogram 5
This relates specifically to the growth stage of a fire, and should only be used for this condition. It therefore applies to the small fire and to the large, but not fully developed fire.

3.3.2 Nomogram 1 – venting a small fire

Enables the vent area for a *small fire* to be calculated. It gives the relation between:

(*a*) the effective depth of the layer of hot gases below the ceiling or roof (i.e. the distance between the bottom of the layer of hot gases and the centre of the vents).

(*b*) the distance between the notional point of origin of the fire and the centre of the vents. (This is equal to $(h_c + 1.5 \sqrt{A_f})$, where h_c is the distance between the floor and the centre of the vents, and A_f is the floor area of the fire.)

(*c*) the square root of the measured free vent area ($\sqrt{A_v}$).

Thus:

if (*a*) and (*b*) are known, then the required vent area (*c*) can be determined;

if (*a*) and (*c*) are known, then the largest fire which those factors will cater for is known, provided the height, floor to vent centre, is known;

if (*b*) and (*c*) are known then the resulting depth of the smoke layer will be determined.

Note: When no units are marked on the Nomograms any units may be used, provided that the same are used for all the scales.

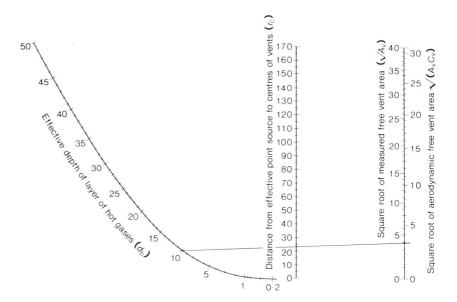

Nomogram 1: *Venting a small fire.*

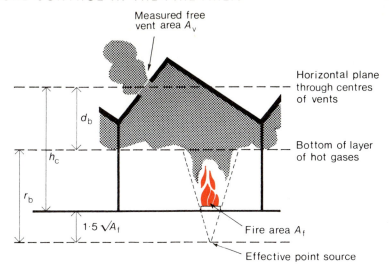

Measured free
vent area A_v

Horizontal plane
through centres
of vents

d_b

Bottom of layer
of hot gases

h_c

r_b

$1 \cdot 5 \sqrt{A_f}$

Fire area A_f

Effective point source

Fig. 3.5   *Notation used in Nomogram for a small fire.*

3.3.3   Nomogram 2 – temperature of layer of hot gases (small fires)

Enables an assessment to be made of the temperature which will apply to the layer of smoke and hot gases which forms underneath a roof or ceiling. It gives the relation between:

(*a*) the distance from the notional point of origin of the fire and the bottom of the layer of hot gases.

(*b*) the heat output of the fire.

(*c*) the temperature of the hot layer.

Heat output of a fire

The use of Nomogram 2 to determine the temperature of the layer of hot gases requires a knowledge of the convective heat output. This will generally be about three quarters of the total heat output, the remaining quarter being dissipated by radiation. An exact estimate of this quantity is difficult to produce. When a fire is burning steadily (or spreading slowly) the burning rate per unit surface of the fuel depends on the nature of the fuel, on the mass, the weight and the disposition of the components of the fuel and to some extent on the shape, size and construction of the enclosure.

Table 3.2 gives some typical values for burning rate per unit surface area of the fuel but these must only be taken as approximate. However, they are sufficient to enable a reasonable estimate of the heat output of a fire to be made for use in Nomogram 2.

Table 3.2   *Heat produced by some fuels*

Material	Calorific value		Burning rate per unit area of fuel surface		Heat output per unit area of fuel surface		See Notes
	Btu/lb	MJ/kg	lb/ft² s	kg/m² s	Btu/ft² s	MJ/m² s	
Wood (also applies for fibre insulating board, hardboard and similar cellulosic materials)	8 000	18.6	1.75 × 10⁻³	8.54 × 10⁻³	14	0.16	1, 2
Petrol	20 000	46.5	7 × 10⁻³	34.2 × 10⁻³	140	1.6	3
Industrial methylated spirit	12 000	27.9	5.5 × 10⁻³	26.8 × 10⁻³	65	0.74	4
Light fuel oil	18 000	41.9	7 × 10⁻³	34.2 × 10⁻³	130	1.5	5

Notes:
1. The calorific value for wood applies for the complete combustion of hard and soft woods.
2. The burning rate for wood is for small cribs in a well ventilated room. For burning with limited ventilation take burning rate = 1/5 mass air flow.
3. Maximum burning rate in a 1.5 ft diameter tray.
4. Steady burning rate in a 1.4 ft diameter tray.
5. Steady burning rate in a deep 9 ft diameter tank.

Examples of calculations which are possible using Nomograms 1 and 2

1. *To determine the total area of vents* (A_v) *needed to keep the layer of hot gases to a given depth* (d_b). For this calculation the following quantities must be known:

d_b = effective depth of layer of hot gases.
A_f = area of the fire.
h_c = effective height of ceiling.

Then calculate the distance between effective point source of heat and the ceiling, from:

$$r_c = h_c + 1.5\sqrt{A_f}$$

Now use Nomogram 1 to obtain $\sqrt{A_v}$ from r_c and d_b and so calculate A_v.

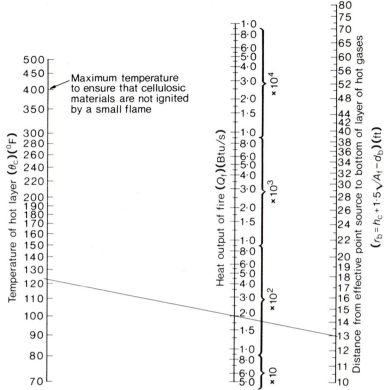

Nomogram 2: *Temperature of layer of hot gases (small fires).*

Conversions:
°F to °C – subtract 32 and multiply by 5/9 : °C to °F – multiply by 9/5 and add 32.
Btu/s to MJ/s – multiply by 1.055 × 10⁻³ : MJ/s to Btu/s – multiply by 948.
Ft to m – multiply by 0.305 : m to ft – multiply by 3.28.

2. *To determine the largest fire from which the hot gases will be limited to a layer of given depth beneath the ceiling.* For this calculation the following must be known:

d_b = effective allowable (or given) depth of hot layer of hot gases.
A_v = total area of vents.
h_c = effective height of ceiling.

Now use Nomogram 1 to obtain r_c from h_c and d_b. Then calculate the area of the fire from:

$$A_f = \left(\frac{r_c - h_c}{1.5}\right)^2$$

3. *To determine the temperature of the layer of hot gases.* For this calculation the following quantities must be known or estimated:

d_b = effective depth of layer of hot gases.

or A_v = total area of vents.

A_f = area of fire.

Q_f = convective heat output of the fire (see Table 3.2).

h_c = effective height of ceiling.

Now calculate position of effective point source of heat from:

$$r_c = h_c + 1.5\sqrt{A_f}$$

and if d_b is not known, use Nomogram 1 to obtain d_b from r_c and $\sqrt{A_v}$.

Now calculate the distance between effective point source and bottom of layer of hot gases (r_b) from:

$$r_b = r_c - d_b$$

Use Nomogram 2 to find the temperature of the hot gases θ_c using the known values of r_b and Q_f.

4. *To determine the total area of vents required to limit the temperature of the layer of hot gases.* For this calculation the following quantities must be known:

θ_c = the temperature required for the layer of hot gases.

A_f = the area of the fire.

Q_f = convective heat output of the fire (see Table 3.2).

h_c = effective height of the ceiling.

Now use Nomogram 2 and find r_b (distance from effective point source to bottom layer of hot gases) using the known values of θ_c and Q_f. Calculate,

$$d_b = 1.5\sqrt{A_f} + h_c - r_b^*$$
$$r_c = 1.5\sqrt{A_f} + h_c$$

Note: If this calculation makes d_b negative, then the layer of hot gases cannot be kept to the temperature θ_c by roof venting.

Then use Nomogram 1 to obtain $\sqrt{A_v}$ from the above values of r_c and r_b.

3.3.4 Nomogram 3 – venting a large but not fully developed fire

This Nomogram determines the relation between the following quantities (Fig. 3.6):

(a) the effective height of the ceiling (measured to the centres of the vents) h_c (in feet);

(b) the effective depth of the layer of hot gases (measured from the centres of the vents), d_b (in feet);

(c) the ratios:

$$\frac{\text{Measured free vent area } (A_v) \text{ ft}^2}{\text{Perimeter of fire } (w_f) \text{ ft}} = F$$

or

$$\frac{\text{Aerodynamic free vent area } (A_v C_v) \text{ ft}^2}{\text{Perimeter of fire } (w_f) \text{ ft}} = G$$

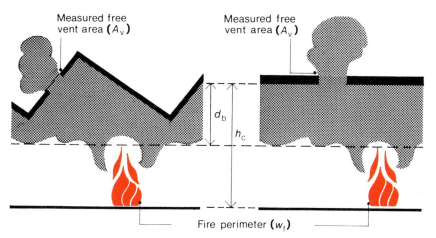

Fig. 3.6 *Diagram showing quantities used in Nomogram 3 for a large, but not fully developed fire.*

The calculations which are possible using Nomogram 3 are:

1. *To determine the total area of vents needed to maintain a given level of the layer of hot gases.* For this calculation the following quantities must be known:

d_b = the required effective depth of the layer of hot gases.
w_f = the perimeter of the fire.
h_c = the effective height of the ceiling.

Now use Nomogram 3 to obtain the value of F using the known values of d_b and h_c. Then calculate area of vent A_v from $A_v = F \times w_f$.

2. *To determine the largest fire from which the hot gases will be limited to a given depth below the ceiling using a given vent area.* For this calculation the following quantities must be known:

d_b = the required effective depth of the layer of hot gases.
A_v = the proposed total vent area.
h_c = the effective height of the ceiling.

Now use Nomogram 3 to obtain the value of the ratio F using the known values of h_c and d_b.

Then calculate the perimeter of the fire, $w_f = A_v/F$, and so find the dimensions of the fire. The length of side of a square fire will be $w_f/4$, or the radius of a circular fire, $w_f/2\pi$.

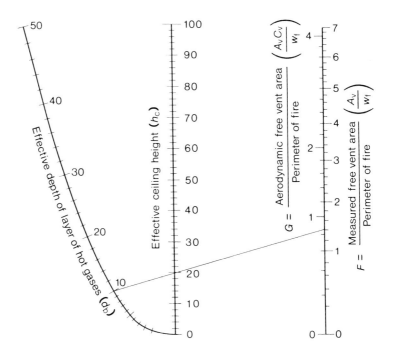

Nomogram 3: *Venting a large, but not fully developed fire.*

3.3.5   Nomograms 4(a) and 4(b) – venting a fully developed fire

Apply to the fully developed fire. When the fire is fully developed it is not possible to control the level of the layer of hot gases because the whole building (or division of the building) is full of fire and smoke. In these circumstances the concern is to prevent the hot gases becoming an agent of fire spread in the building by ensuring that all the smoke and hot gases go out of the building at roof level and that at the lower openings only fresh air is flowing into the fire area. The *neutral plane* concept is used in the two Nomograms. This can be described as the horizontal plane in a fire room, above which the pressure is above ambient and will, therefore, act to drive the smoke and hot gases out of the room; and below which the pressure is below ambient and so will act to draw air into the fire area. At the level of the neutral plane the pressure in the fire room will be equal to ambient and consequently, there will be no flow into or out of the fire room at this level.

The Nomograms 4(a) and 4(b) relate the factors which control the position of the neutral plane and so enables the calculation of the vent area necessary to ensure that the neutral plane is so placed that no hot gases can flow out of the low level openings. *It should be emphasized that the two Nomograms can only be used when the total area of low-level openings is less than 1/20 of the area of the fire.* The notation used is shown in Fig. 3.7. The quantities used in the Nomograms are:

B = stack height [distance between the neutral plane and the effective ceiling (which is the centre of the vent areas)].

P = distance between the neutral plane and the top of each low level opening.

X = the width of each low level opening.

Y = the height of each low level opening.

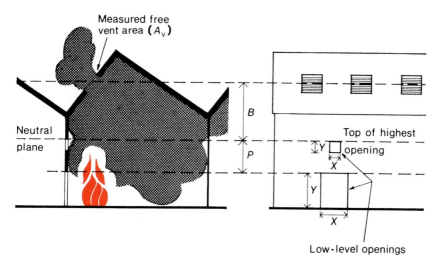

Fig. 3.7   *Notation used in Nomograms for venting a fully developed fire.*

The first step in the calculation is to *use Nomogram 4(a)* to obtain a characteristic parameter (N) for each low level opening. This parameter depends on the size and position of each opening and to find it, using Nomogram 4(a), it is necessary to know:

Y = height of each opening
X = width of each opening
P = distance between the top of each opening and the required position of the neutral plane

and hence

P/Y = position factor for each opening.

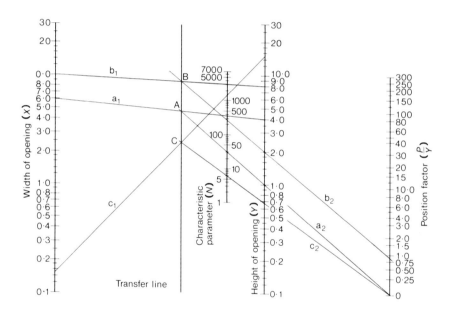

Nomogram 4:   *(a) Venting a fully developed fire.*

The required position of the neutral plane will be level with, or just above, the top edge of the highest opening.

To use the Nomogram, draw a line between the appropriate positions on the width scale (X) and on the height scale (Y). Mark where this line crosses the *transfer line* – now join this point on the *transfer line* to the appropriate value of P/Y on the right-hand scale. Read off where this line crosses the characteristic parameter (N) scale.

Do this for each of the low level openings and add together all the values of *N* so obtained. This gives Σ*N*.

Now use Nomogram 4(b): On the left-hand scale mark off value for *B*, which is the distance between the neutral plane and the centre point of the roof vent. On the centre scale mark off the value of Σ*N* obtained as above. Join these two points and produce the line to cross the right-hand scale and so read off the appropriate vent area. (The scale gives the square root of the vent area.) In making the calculation of the vent areas an allowance may be necessary for the general air leakage through the walls of the building.

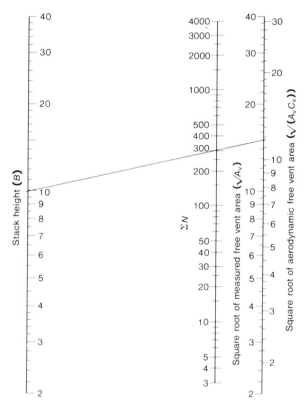

Nomogram 4:   *(b) Venting a fully developed fire (continued).*

Example of the use of Nomograms 4(a) and 4(b)

A building has a flat roof 25 ft high. Part of the building is separated off by a fire-resistant brick wall (Fig. 3.8). The size of this part is 50 ft × 50 ft. The separating wall has two openings in it: a door opening at ground level 10 ft wide × 8 ft high, and a conveyor (or services) opening 4 ft high and 6 ft wide, the top of which is 15 ft above the ground. There are no openings in the outside walls but the leakage is estimated to be 0.1 per cent of the wall area.

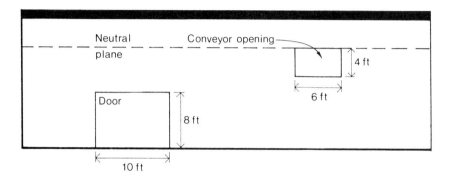

Fig. 3.8   *Diagram of a dividing wall.*

The position of the neutral plane must be level with the top of the conveyor opening, i.e. 15 ft above the floor. It is required to find the total vent area which will maintain the point at this level:

1. *Consider the three low level openings and find ΣN. Thus,*
   Opening 1 (conveyor opening):

   $X = 4$ ft (width)
   $Y = 6$ ft (height)
   $P = 0$ (because neutral plane is level with top of opening).
   So, $P/Y = 0$

Draw lines a_1 and a_2 on Nomogram 4(a) and so find that $N_1 = 30$ for this opening.
   Opening 2 (door):

   $X = 10$ ft (width)
   $Y = 8$ ft (height)
   $P = 7$ ft (distance between top of opening and neutral plane).
   So, $P/Y = 7/8 = 0.875$

Draw lines b_1 and b_2 on Nomogram 4(a) and so find that $N_2 = 270$ for this opening.

Opening 3 (the leakage), is estimated to be 0.1% of outside wall area. Since the leakage is evenly distributed it is considered for this purpose to be equivalent to an opening in the wall which extends from the neutral plane to the ground; and since there are three outside walls each 50 ft long the width of this opening would be 0.15 ft.

$X = 0.15$ ft (width)
$Y = 15$ ft (height)
$P = 0$ (top of opening level with neutral plane)
$P/Y = 0$

Now draw lines c_1 and c_2 on Nomogram 4(a) and so find that N_3 for this leakage opening $= 6$. Then,

$$N = N_1 + N_2 + N_3 = 306$$

2. *Determine the distance between the centre of the roof vent and the neutral plane (B).* In this example the ceiling is flat and so the centre of the vent coincides with the ceiling. Thus $B = 10$ ft (height of ceiling $= 25$ ft, height of neutral plane 15 ft).

3. *Use Nomogram 4(b), and with the above values of ΣN and B find that the square root of the required vent area $= 15.1$.* Hence, the required total vent area (measured) is $(15.1)^2 = 228$ ft^2.

3.3.6 Nomogram 5 – height of flames from a wood fire

This Nomogram gives the relation between the following three quantities:

(a) the flame height (L ft)
(b) the convective heat output per unit area of the fire (Q_f/A_f in Btu ft^{-2} s^{-1}). This is roughly three-quarters of the total heat output per unit area.
(c) The mean diameter of the fire (D ft). This may also be the length of the side of a square fire.

The relation between these three factors will enable, in some cases, an estimate to be made of the convective heat output of a fire, a quantity required in the use of Nomogram 2. An estimate of the flame height, on the other hand, will assist in the determination of the temperature likely to apply to the layer of hot gases under a ceiling.

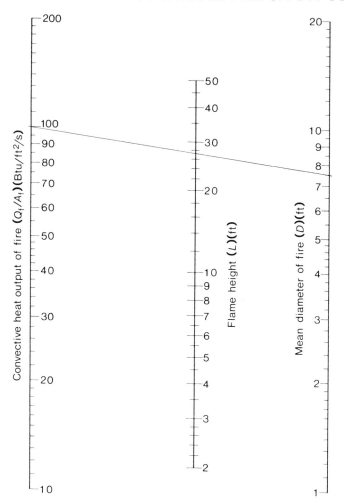

Nomogram 5: *Height of flames from a wood fire.*

3.3.7 Nomogram 6 – fraction of heat output of the fire exhausted by the vent

This Nomogram relates the following four quantities, using two straight edges and the transfer line as shown on the scales:

(*a*) the depth of the layer of hot gases (d_b).

(*b*) the distance from the effective point source to roof screen ($r_g + h_c$).

(c) The fraction of heat vented (Q_v/Q_f).

(d) The square root of measured free vent area. (A discharge coefficient of 0.6 is assumed. Where this is not the case, due perhaps to a restricted inlet, a correction must be made.)

This Nomogram applies to a small fire and enables an assessment to be made of how effective a given vent design will be in releasing the heat developed by a possible fire.

No dimensions are shown on the Nomogram. Q_v/Q_f is dimensionless. Any units may be used for the other scales provided the same are used for all of them. The range of the Nomogram may be extended by multiplying all the length scales by the same factor.

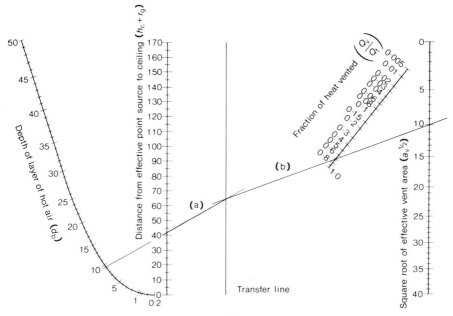

Nomogram 6: *Fraction of the heat output of the fire exhausted by the vent.*

3.4 OTHER RELEVANT FACTORS

3.4.1 Temperature of the layer of hot gases

It may be important to estimate the temperature of the layer of hot gases so that some idea can be formed as to the effect such a layer could have on the material used in the construction of the building; for instance, if unprotected steel is present or if the ceiling has any combustible material used in its make-up.

TABLE 3.3 *Temperature of layer of hot gases*

Convective heat output of fire per unit area.	Area of fire	Height of clear layer (metres)	Temperature of hot layer (above ambient) °C
1 MW/m²	3 m × 3 m	10	82
(88 Btu/ft² s)		7	143
		5	227
		4	> 300
	2 m × 2 m	10	52
		7	96
		5	160
		4	232
		3	> 300
	1 m × 1 m	10	10
		7	35
		5	68
		4	104
		3	173
200 kW/m²	3 m × 3 m	10	13
(17.5 Btu/ft² s)		7	29
		5	49
		4	65
		3	90
	2 m × 2 m	10	< 10
		7	15
		5	32
		4	46
		3	65
	1 m × 1 m	5	< 10
		4	15
		3	38

As long as the fire is small and the bottom of the layer of hot gases is well above the tips of the flames, then *the temperature of the layer* will depend on *the heat output of the fire* and the distance between the bottom of the layer and the effective point source of the fire. The temperature of the layer (provided it is below about 300°C (500°F)) may be calculated using Nomogram 2 (page 50) but a few examples are given in Table 3.3.

The temperatures listed in Table 3.3 (and those determined by the use of Nomogram 2) are deduced by assessing the volume of cold air entrained into the smoke plume. When the heat output from the fire has an appreciable radiation component (i.e. for the larger fires listed) and when the ceiling is constructed of material which is a good thermal insulator, the temperature of the ceiling surface may be higher than the values given.

3.4.2   Inlets for cold air

Venting will not be effective unless it is possible for cold air to flow into the building to replace the hot gas which is required to flow out through the roof vents. Thus, adequate inlets for cold air must be provided. If the total area of the cold air inlet is too small then the roof vents will behave as if their size was reduced. Fig. 3.9 shows how the area of the cold air inlets affects the effectiveness of the smoke vents. For the effectiveness of the smoke vent to be at least 90% of its calculated value the ratio of cold air inlet area to smoke vent area must be:

(*a*) over 2 if the layer of gases beneath ceiling is cold.
(*b*) $1\frac{1}{2}$ if the layer of gases is at 250°C above ambient.
(*c*) 1 if the layer of gases is at 800°C above ambient.

The likely temperature of the layer of hot gases must be assessed for the fire size, building dimensions, and smoke level required and the air inlet size arranged accordingly. Ideally, these inlets should be situated as close to the ground as possible so that they are well below the level of the bottom of the layer of hot gases. This is necessary to prevent the cold air flowing into the building entraining smoke from the layer and so producing smoke logging at ground level.

It is preferable to arrange that these inlets are well distributed round the building. In the early stages of a fire when only one or two roof vents are open, the natural leakage of air in through doors and windows may be sufficient, but as the fire develops more air inlet area will be needed. This can, if necessary, be provided by automatically or manually operated openings.

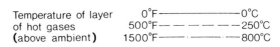

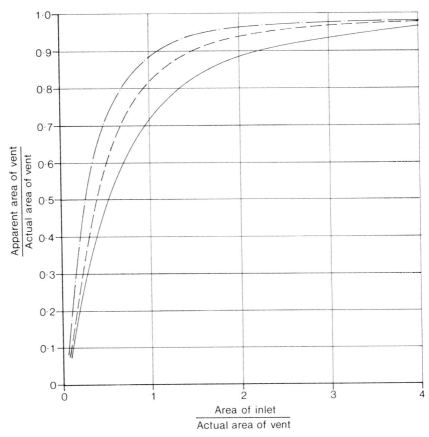

Fig. 3.9 *Effect of area of cold air inlet on apparent area of smoke vent.*

3.4.3 Size of individual roof vents

Although all the design calculations are directed to determining the area of roof vent required in any given circumstances, it must be understood that this relates to a total area and that in general the venting should be accomplished by using several small vents rather than a single large vent (Fig. 3.10).

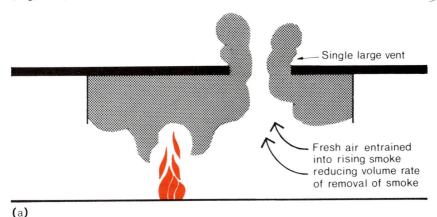

Single large vent

Fresh air entrained into rising smoke reducing volume rate of removal of smoke

(a)

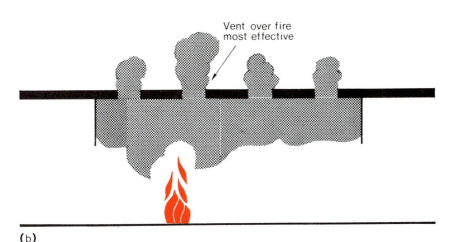

Vent over fire most effective

(b)

Fig. 3.10 *(a) Large vent is less efficient. (b) Several small vents are more efficient.*

There are several reasons for this:

(a) If the vent is large, that is to say, its dimensions are similar to the depth of the smoke layer, then the flow of gases out will disturb the bottom of the

smoke layer and air will be entrained into the rising smoke, thus reducing the effectiveness of the extraction.

(*b*) Vents directly over a fire are more efficient and more effective than those some distance from the fire. The gases are hotter and so escape more quickly, and thus the vent allows a greater volume of hot gas to escape. However, since the exact position of the fire is not known it is preferable to distribute the vents over the whole possible fire area so that at least one or two will be close to any fire and so will open quickly and be effective in allowing smoke to escape at an early stage of the fire.

(*c*) If the fire develops to the stage that flames emerge from the vent, then the height of emergent flames will be less with a small vent than with a large opening so that the exposure hazard to external roofing material or adjacent buildings will be reduced.

3.4.4 Restricting lateral spread by dividing the space under the ceiling or roof

In general discussions about the design of roof venting systems the idea of establishing smoke reservoirs below the ceiling or roof is suggested. In small buildings this may not be necessary, but in large buildings it is desirable to restrict the sideways spread of hot gases beneath a ceiling. There are two reasons for this:

(*a*) If the temperature of the hot gases is high, then it is obviously desirable to limit the area over which they can spread and cause damage or even ignition to combustible parts of the ceiling lining or service components.

(*b*) Even if the temperature is not particularly high then in the course of sideways travel the smoke layer will (because of mixing with cold air) become relatively cool and shallow. This will result in any roof vents being inefficient because a vent will be most effective if the temperature of the gas is high and if the layer beneath the vent is quite deep, thus developing a reasonable pressure differential to act to expel the hot gases through the vent.

The dimensions of any sub-division of the area under a ceiling into smoke reservoirs will depend on three factors, which are:

(*a*) the size of fire;
(*b*) the depth of the layer of hot gases;
(*c*) the height of the building.

For any given building, factors (*b*) and (*c*) are probably fixed and so the critical factor is the *size of fire*. However, in all the design considerations relating to roof venting, an assumption has to be made as to the likely size of fire. Consequently, it is possible to assess the correct size of the smoke reservoirs by using the criteria already established for the roof vent design.

Using the Nomograms provided in this Chapter, it is possible to calculate the maximum size of fire whose hot gases can be confined to any given size smoke reservoir, and examples of these sizes are given in Table 3.4. However, in the absence of sufficient design data to accurately assess the smoke reservoir size it is suggested that 60 m × 60 m be treated as the maximum size smoke reservoir which should be used.

Requirement for screens which form smoke reservoirs

Screens which extend downwards from the roof or ceiling to form the smoke reservoir must be non-combustible and gas tight. They should preferably be as resistant to the effects of the fire as the roof members. Small leaks or gaps in the screen (for instance, where pipes or services pass through) are not of great importance provided they are as small as possible and not numerous.

Screens should always be as deep as possible because this improves the effectiveness of the venting and enables a larger fire to be confined. It also provides the maximum possible delay before smoke spills into adjacent compartments. It may be advantageous to direct the smoke spilling out of any smoke reservoir away from areas of risk or to avoid smoke damage. This can be done by providing deeper screens on the sides of the reservoir facing the areas which it is required to avoid.

3.4.5 Effect of adverse wind conditions

When considering the position of the roof vents and of the cold air inlets some thought must be given to the possible effect of external wind conditions. The pressure developed by a wind may well be greater than that which is causing the hot gases to move out of a roof vent and this could result in cold air entering at roof level carrying with it the smoke which it was intended to release. This will then be carried downwards in the building and low level smoke logging could occur.

The pressure at a roof vent due to a wind will depend on the aspect of the vent opening, on the configuration of the roof and on the position of the low level air inlets. If the roof vent discharges vertically, or nearly vertically, then the effect of wind blowing across a roof will usually result in a suction which will assist the action of the roof vent. However, with roofs which have a steep pitch (greater than about 40°) the pressure on the windward slope will tend to oppose the flow of hot gases through the vent if the plane of discharge is parallel to the roof surface. Whenever possible, with large buildings having north light roofs, vents should not be placed in the outward facing steep roof

TABLE 3.4: Maximum size of side of square fire which can be confined to a single roof compartment

Max. size of side (in metres) of square fire for smoke and hot gases to be contained by vents of size (% of reservoir area)

Area of reservoir	Height of building	0.5%			1.0%			2%			5%		
		With smoke curtains of depth:											
		1 m	2 m	3 m	1 m	2 m	3 m	1 m	2 m	3 m	1 m	2 m	3 m
(20 m × 20 m) 400 m²	3 m	1.3	4.1	—	2.5	8.2	—	5.1	16.5	—	13	41*	—
	6 m	0.3	0.5	1.2	0.6	1.0	2.4	1.3	1.9	4.7	3.2	4.8	12
	9 m	0.2	0.3	0.4	0.3	0.5	0.8	0.6	1.1	1.6	1.6	2.7	4.0
(40 m × 40 m) 1600 m²	3 m	5.1	16.5	—	10	33	—	20.3	66*	—	51*	165*	—
	6 m	1.3	1.9	4.7	2.6	3.9	9.5	5.2	7.7	19	13	19	47*
	9 m	0.7	1.1	1.6	1.3	2.1	3.2	2.6	4.2	6.4	6.6	10.6	16
(60 m × 60 m) 3600 m²	3 m	11	37	—	23	74*	—	46*	148*	—	114*	370*	—
	6 m	3	4.4	10.6	5.8	8.7	21	12	17	42*	29	44*	106*
	9 m	1.5	2.4	3.6	2.9	4.8	7.2	5.9	9.6	14	15	24	36

* *Note*: the theory is only approximate for fire size greater than 40 m × 40 m.

slope because of the unfavourable action of a wind. If this cannot be avoided, then consideration should be given to the use of vents which are so designed that the wind will always produce a suction effect.

Wind blowing into low level openings will tend to create a pressure in the building which will assist the hot gases to move out of the roof vents. Similarly openings on the leeward side may result in a suction effect and so oppose the action of the roof vents. For this reason the low level air inlet openings should be distributed round the building and there may be some merit in these openings being under the control of responsible persons who can then assure that only those on the windward side of a building are open.

3.4.6   The importance of early venting

The venting of a fire in a building can be achieved by several means: firemen breaching the roof; opening roof vents manually, or by the fire operating fusible links controlling the vents; or by softening plastic roof lights which then fall out. The choice of method may depend on the individual circumstances of a particular building, but whatever the building the earlier the venting takes place the better, so that in general terms there is a very strong case for using automatically opening vents.

The important features of early venting are:

(*a*) It may be much easier to prevent a building becoming smoke logged than to clear smoke from it once it has become completely filled with smoke. This is because in the latter stages of the fire, the smoke movement in it will become sluggish and even stagnant, and opening vents at this stage will be less effective than earlier when the upward flow of smoke and hot gases is still well established.

(*b*) When a fire in an unvented building is large or has been burning for some time, it may become partially starved and unburnt flammable gases may collect beneath the ceiling and in the smoke reservoirs. If venting then occurs there may be energetic ignition of these gases with dangerous possibilities. Early venting will prevent the accumulation of such flammable products of incomplete combustion.

(*c*) If the fire damage is to be confined to a single roof compartment, then the roof vents in that compartment must open before it becomes full of smoke and spillage of smoke and hot gases to adjacent reservoirs takes place.

(*d*) Early venting may lead to early detection of a fire in an unoccupied building and will minimize the spread of the fire because of the release of heat.

If reliance is placed on the manual opening of roof vents, this may involve the attempt to locate low level control points in a smoke logged building, or a remote control arrangement may have been rendered unusable by the fire before the opening action has been completed.

3.4.7   Roof vents and sprinklers

Roof vents must not be regarded as being an alternative to sprinklers. Roof vents are a smoke control feature but sprinklers have an extinguishing function. There may well be sound arguments for using both roof vents and sprinklers, for instance, even if the sprinkler system extinguishes the fire, a large quantity of smoke will have been produced before this is achieved and the release of this smoke (and the heat in it) must be an advantage. On the other hand, the sprinkler system will have fulfilled its purpose if it controls the spread of fire and in this case the release of smoke produced could well become essential.

Flames spreading under a ceiling may pass through a sprinkler discharge without being noticeably cooled. This may result in the operation of sprinkler heads away from the floor area affected by fire, but the use of roof screens and vents would limit the sideways spread of smoke and flame and prevent the unnecessary opening of remote sprinkler heads.

The use of roof screens will tend to hasten the opening of the affected sprinkler heads, but the opening of a roof vent will initially reduce the temperature beneath the ceiling and may, therefore, act to delay the operation of the sprinkler heads.   For this reason it is always suggested that when sprinklers and automatically opening roof vents are both used in a building the sprinkler head should be designed to operate before the roof vents open. At the same time, however, the possibility of the sprinkler discharge falling on the fusible links which operate the roof vent must be prevented; if necessary by the use of suitably placed screens protecting these fusible links from the water spray.

REFERENCES

1. Thomas, P. H., Hinkley, P. L., Theobald, C. R. and Simms, D. L. (1963). Investigations into the flow of hot gases in roof venting. Fire Research Technical Paper No. 7, London HMSO.
2. Thomas, P. H. and Hinkley, P. L. (1964). Design of roof venting systems for single storey buildings. Fire Research Technical Paper No. 10, London HMSO.
3. Hinkley, P. L. and Theobald, C. R. (1966). P.V.C. roof lights for venting fires in single storey buildings. Fire Research Technical Paper No. 14, London HMSO.
4. Yokoi. S. (1960). A study on the prevention of fire spread caused by hot upward current. Japanese Ministry of Construction Building Research Institute Report No. 34, Tokyo.

4 Smoke control in escape routes without complete structural protection – covered shopping areas

Traditionally, planning for the safety of life in buildings has followed a fixed pattern generally providing:

(*a*) that the rooms where fires could occur are kept as small as possible and are structurally self-contained to defined standards of fire resistant construction.

(*b*) that the travel distances across these floor areas by which people can reach a 'protected point' are within prescribed distances.

(*c*) that the 'protected point', i.e. a door leading to a 'protected route', is wide enough to enable the anticipated number of people on that floor to enter within a prescribed time.

(*d*) that the 'protected route' has prescribed standards of fire resistant construction, is totally self-contained and leads from all the floor area at risk directly to ground floor level, into open air, taking people away from the building.

In recent years changes in types of buildings and the ways in which different buildings are assembled, have necessitated a complete rethinking of these traditional concepts, and a classical example is the enclosed shopping centre. Large numbers of people will be walking around the shopping malls and therefore not strictly speaking within 'buildings', which are the shops and department stores, theatres and restaurants, etc. The application of Building Regulations and statutory fire precaution requirements directly applies to a building and when many buildings of many different occupancies are grouped together it may be possible to apply the traditional concepts in planning the means of escape, but these need not apply to the common enclosing parts. However, these common parts will contain a great many people who could be at risk or might panic if a fire broke out. It has been demonstrated that the risk of fire within these common parts is generally small and indeed restrictions are usually placed on the possible combustible elements or contents within these areas.

If a fire occurs in any one of the buildings comprising such a complex, however, the smoke and hot gases are quickly spread from the building into these common parts and a great deal of thought has been given to designing smoke control systems for such areas, to ensure that smoke penetration is rapidly exhausted to atmosphere, and that its layering at ceiling level is contained well above head height so that free vision as well as free passage is maintained for anybody trying to escape in these areas.

It may be difficult to visualize the extent of smoke spread and the type of damage that can develop from a relatively small fire; the case history of the fire at the combined Wulfrun and Mander Shopping Centres in Wolverhampton (1970) gives a clear indication of these problems.

4.1   CASE HISTORY – THE WULFRUN AND MANDER SHOPPING CENTRES

4.1.1   The buildings

The Wulfrun and Mander shopping centres together cover an area of about 34 acres (14 ha) of compact shopping precinct providing some 220 rentable shop units of various sizes. The buildings were planned and started before the advent of the Building Regulations. The two centres are linked by a covered bridge allowing pedestrians to walk from one to the other (see Fig. 4.1). The precincts are on various levels, with entrances from all the streets adjoining the development, and include a 10 storey tower office block and two multi-storey car parks. An extensive basement provides service and loading areas for all units.

4.1.2 Origins of the fire

The fire started in a retail carpet shop in the Wolfrun Centre (No. 1, Fig, 4.1) at some time between 18.45 h the evening before it was discovered and 05.40 h on the morning of the fire, when a burning smell was noticed by someone 725 ft (221 m) away. Thinking that some rubbish was being burned this person took no action. Someone else delivering bread at 05.15 h to a shop two doors away did not see anything unusual except that there were no display lights on in the carpet shop. It is possible, however that smoke and unburned gases had already percolated through the ventilation openings over the carpet shop display windows and entered the large void above the suspended decorative ceilings in the arcade outside the shop.

A passer-by saw smoke near a large store in the Mander Centre, 400 ft

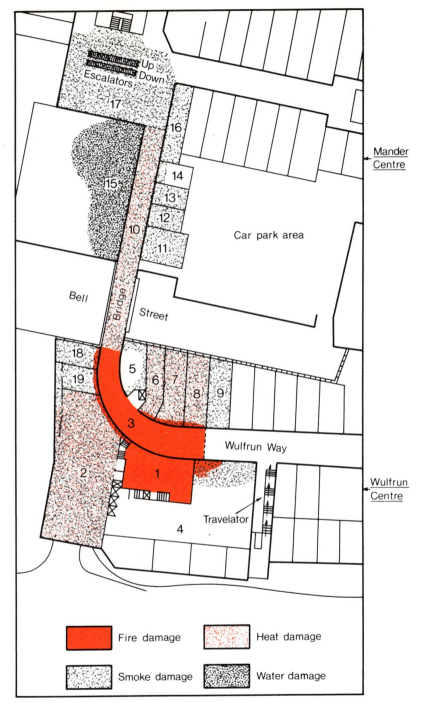

Fig. 4.1 *Plan of the Wulfrun and Mander Shopping Centres.*

(122 m) from the seat of the fire and at 06.08 h called the fire brigade who were in immediate attendance. The plate glass windows of the carpet shop fractured and the fire burst into the arcade with some violence, both fire and very considerable heat passing rapidly down the covered Wolfrun Way in the direction of the Mander Centre. At 06.12 h five sprinkler heads operated just outside and three heads inside a large store 376 ft (115 m) away from the fire. The heat then dispersed to the open air at Mander Square (top left-hand side of Fig. 4.1). The store sprinkler system was directly connected to the fire brigade control room by GPO line. At 06.15 h the light-sensitive burglar alarm in another shop more than 370 ft (113 m) away from the fire was actuated by smoke entering the shop. By this time, however, Wolverhampton fire brigade appliances were working at the scene. When they arrived, firemen found a very severe fire spreading rapidly down the enclosed mall and beginning to involve ten shops. A brisk wind was encouraging the spread.

4.1.3 Fire development

The pattern of damage showed the rapid heat spread along considerable stretches of the shopping mall; fire had travelled in a spiral motion causing an uneven pattern of damage. Fire spread was encouraged by the combustible ceiling linings in the mall; smoke logging and the lack of means of venting the fire presented firemen with a difficult situation. The pedestrian way was 28 ft (8.5 m) wide at the carpet store and after turning in an arc through 90° it narrowed to 18 ft (5.5 m). As the fire travelled along the mall it affected the shops on each side, the amount of damage diminishing with distance, but it was still sufficiently intense to operate the sprinkler heads already mentioned 376 ft (115 m) away.

4.1.4 Damage

In the carpet store itself all the stock and fittings were severely damaged by fire and there was considerable spalling of the concrete ceilings and columns. The basement areas were slightly damaged by water. Four premises nearest to the carpet store (a wine and spirit merchant, a fresh food shop, a butcher and a firm selling educational products, books and handicrafts) suffered severe fire damage to the shop frontage and window stocks with severe heat, smoke and water damage to the rest of the decor and contents. At basement and mezzanine levels there was slight damage by heat, smoke and water.

Shop window fascias of four more shops in the Wolfrun Centre were damaged by fire and heat and the rest of the contents severely damaged by smoke. This pattern of damage recurred, though not quite as severely, in the Mander Centre on the other side of the covered bridge where four more shops were damaged by smoke. A department store was slightly damaged by smoke and also by water from a sprinkler system. Part of the concrete ceiling to the covered way was spalled and the decorative timber roof lining was damaged in the shopping mall itself for a distance of 380 ft (116 m) in both shopping centres but this was less severe in the Mander Centre [22].

4.2   THE DESIGN OF ESCAPE ROUTES

The normal philosophy regarding the design of escape routes in buildings is that an occupant should be able to move away from the fire across the floor of the compartment containing it, and reach a fire-resisting door which leads into a protected escape route (protected, that is, by fire-resisting construction – walls, floors and ceiling). Once inside, the occupant should be safe from the effects of the fire for the prescribed time. However, in some buildings, notably in covered shopping complexes, the main escape routes cannot, for functional reasons, have this fire-resisting protection. For instance, the shops will be ranged along each side of the pedestrian malls and the shop fronts may not even be glass windows – in some cases completely open fronts are preferred for commercial reasons.

Nevertheless, the pedestrian malls are the main escape routes for the occupants of the shopping centre, and there may well be several thousand people at risk. Consequently, the malls must be maintained in a usable state for those people to escape, and the escape movement of such a large number may well take a considerable time. Effective control of the smoke leaking into the pedestrian malls is, therefore, a vital feature of the design considerations of such buildings.

Information for use in the design of smoke control systems in this type of building has, until recently, been limited but is now becoming available. Research on the various problems is still being carried out and design experience is being accumulated as new shopping centres are designed. However, the general principles relating to smoke venting as described in Chapter 3 are adopted and applied.

There is a Fire Prevention Guide [1] published by the Home Office which deals with fire precautions in town centre redevelopment, and this makes suggestions relating to smoke control applicable to single storey malls. The information in this Guide is based on work carried out at the Fire Research Station of the Building Research Establishment (at Borehamwood, England) and there are several publications describing this work [2–11]. These are valuable documents and should be consulted to complement the information given below, some of which has been drawn from these sources.

There are two different ways of approaching the problem of smoke control in the type of building being considered here. These are:

(*a*) To install a smoke control system in all the shop areas. This will be designed to ensure that no smoke can escape into the pedestrian malls and if practicable would be considered the ideal solution.

(*b*) To accept the idea that smoke from a fire in a shop will spread into the pedestrian malls and to install smoke control in these malls to ensure that the smoke will not travel far along the mall and remain at high level so that pedestrians can safely move about below it.

It may not always be possible to treat the approach as being one or other of the above. A combination of the two systems may sometimes be considered, but the design rules described below will still apply.

4.2.1 Buoyant smoke and stagnant zones

The general idea for smoke control systems described in this Chapter is that the smoke from a fire in a shop will be hot enough, and, therefore, buoyant enough to form a stable layer under the ceiling without mixing with the clear air at low level. At the same time, the smoke will be removed from this layer either by venting or by extraction and it is important to avoid, if possible, the formation of stagnant zones either in the smoke layer or in the clear air beneath. These can arise when venting or extraction is not properly distributed or when a mall has a closed end.

If a smoke layer is stagnant, it will cool down and is then likely to mix into the air below. If the air space under a smoke layer is stagnant, it will tend to fill up with hazy smoke of sufficient density to affect visibility. This effect is due to a variety of reasons; for instance, the stagnant air will be quickly warmed up at the smoke–air interface, and allow mixing with smoke (particularly if the smoke above is stagnant); or smoke may creep down the side or end walls from that part of the smoke layer which is in contact with (and cooled by) the walls [9].

It should, therefore, be a design feature that:

(*a*) venting or extraction should be well distributed in the smoke layer or smoke reservoir; and

(*b*) the air inlets at low level should also be well distributed so that the entering air will purge as much of the low-level clear air as possible.

If in practice these features cannot be ensured, then some attempt should be made to introduce secondary extraction, or in the second case, air supply into zones which could be possible dead-end situations. It is probable that the normal ventilation system can be used for this purpose.

4.2.2 Effect of sprinklers on smoke layering

It has been shown in Chapter 1 that the quantity of smoke produced in a fire is very large, and it has been stated in Chapter 2 that in designing a smoke control system an assumption has to be made about the size of a possible fire. In the buildings being considered here, the escape route must be kept usable for a long period, and this means that the fire size must be limited so that there is no question of the smoke control installation becoming overwhelmed by a growing fire. It is essential, therefore, that sprinklers be installed in all areas where a fire can occur. This includes the malls if combustibles are likely to be placed in them at any stage of the normal operation of the shopping centre.

Since sprinklers are essential in order to limit the size of a possible fire, some consideration must be given to any likely interaction between smoke layering and sprinkler operation.

There are two ways in which the sprinklers can affect the smoke:

(a) They can, by the discharge of water spray through the smoke layer, bring the smoke down to low level; and

(b) By cooling the smoke they can reduce its buoyancy and so make it move more slowly through the roof vents.

The interaction between sprinklers and smoke has been noted in a number of tests without any clear pattern emerging.

In 'Operation School Burning' [12], a series of tests carried out by the Los Angeles Fire Department, it is recorded that sprinklers reduced the quantity of smoke produced by acting on the burning material. In contrast, other experiments [13] have suggested that smoke-logging at low levels is caused by sprinkler operation. In work carried out at the Fire Research Station, Borehamwood (referred to in Chapter 3) it is recorded that in some circumstances, although sprinklers carried smoke down to a low level, it was subsequently drawn back into the fire with the entrained air and except in the immediate vicinity of the fire the smoke problem was not increased by sprinkler action.

Because of the importance, for smoke control purposes, of confining the smoke to a stable high level layer, a closer look at the action of sprinklers has been reported by Bullen [14]. He reports that for thick smoke layers (i.e. of 1 m or more), if the layer is hot enough to set off the sprinklers then it will be buoyant enough to remain as a layer even when the sprinklers are operating. As the thermal and smoke output of the fire is reduced by the sprinkler action, the smoke layer in the shop or in the mall will become cooler and smoke may then be dragged down to a low level, but by the time this stage is reached all the occupants should have been able to escape. It may be a problem for the Fire Brigade, however, but by now the fire should be under control.

Recent (and unpublished) calculations of the cooling of smoke layers moving away from a 5 MW fire (i.e. the size of fire usually assumed for a sprinklered condition) show a rapid decline in temperature, and generally a fire in a large shop or store would give off smoke and gases in the mall which are too cool to set off a sprinkler; but for small shops the same size fire (5 MW) would be expected to set off several sprinkler heads in the mall.

In a continuation of the work, reported by Morgan [15], a study has been made of the effect of sprinklers on a system which relies on the natural venting of smoke from smoke reservoirs. Difficulty arises because the sprinkler action will cool the smoke in the reservoir, render it less buoyant, and so affect the efficiency of the natural vents provided. In an example worked out to illustrate his work, Morgan suggests that the vent area from a smoke reservoir in which sprinklers could cause cooling needs to be increased by 27% in order to keep the smoke at the predetermined level.

It is difficult to generalize from this one example, but there is no doubt about the reality of the effect and it is, therefore, suggested that in smoke reservoirs which are relying on natural vents for smoke removal, and where these reservoirs also contain sprinklers, the total vent area should be increased by 20–25%. Where the smoke removal is by powered extraction there is no reason to make this allowance; in fact, the cooling of the smoke before it enters the extract system must be advantageous.

4.3 DESIGN DETAILS FOR SHOPS FRONTING ON TO COVERED PEDESTRIAN MALLS

4.3.1 Smoke confined to the shop of origin (Fig. 4.2)

The simplest situation under this heading is the case when the shop front is of fire-resisting construction and is fitted with self-closing fire-resisting doors. For functional reasons, this situation is unlikely to apply, and even if it were possible to assume that structural containment of the smoke would be complete, it would still be advisable to consider smoke control in the shop in order to keep the smoke at a high level, and so create safe conditions for the escaping occupants, who could be numerous and possibly unfamiliar with the layout of the building.

The purpose of smoke control in the shop area is to keep the smoke confined in its lateral spread and to keep the lower surface of the layer of hot gases and smoke well above the heads of the escaping occupants. The level chosen for this lower surface will depend on the overall height of the shop, but it should always be at least 2.5 m above the floor.

In this connection a distinction between 'small' and 'large' shops or stores needs to be made. In the latter case it is not only feasible to retain the smoke within the limits of the shop, but it is also highly desirable. The reason for this is that if the smoke fills a large shop before entering the mall, then considerable cooling will have taken place and the smoke may be too cool to be disposed of by the mall venting or extraction system and low-level smoke-logging of the mall could result. It is therefore suggested that when shops front on to an extensive mall system they should have their own smoke control system if they exceed 1000 m² in floor area. This requirement is to provide protection to the people in the malls as well as assisting the occupants of the affected shop to escape. For a small shop, below 1000 m², the provision of an individual smoke control system may not be feasible on account of the size of extraction required.

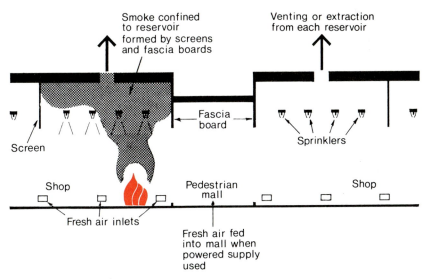

Fig. 4.2 *Smoke confined to the shop of origin.*

The requirements in the design of smoke control system for shop areas are as follows:

(*a*) Sprinklers must be installed in the shops so that the fire size can be assumed to be limited to 3 m × 3 m. (This gives a fire perimeter of 12 m and it is further assumed that the heat output of this size fire will be 5 MW. This assumption is valid for most merchandise to be found in shops but it might need to be reconsidered if concentrations of very combustible and rapidly burning material is involved.).

(*b*) The space underneath the ceiling must be divided into 'smoke reservoirs' by the use of screens extending downwards from the ceiling. These screens should be as deep as possible, consistent with their lower edge being at least 2.5 m from the floor. In any case they should be at least 1 m deep. The maximum area for the smoke reservoirs should be taken to be 1000 m². If the shop area is no greater than this, then the whole shop can be considered to be the smoke reservoir. This size limit is set to avoid too much cooling of the smoke layer which might then impair the efficiency of the venting systems. If mechanical extraction of the smoke is proposed this size limit for the smoke reservoir need not be rigidly adhered to, but even so the size would not be greater than 1300 m².

If the construction incorporates a false ceiling, then the space above the false ceiling may be included in the depth of the smoke reservoir provided the screens forming the reservoir extend up to the structural ceiling, the perforations in the false ceiling are at least 40% of the total area, and the space occupied by services etc. above the ceiling is not greater than 50% of the volume above the ceiling. If the perforations are greater than 40%, then provided there is sufficient depth available, the whole of the smoke reservoir may be above the false ceiling and the screens forming the smoke reservoir need not project below it.

At the shop front where it adjoins the pedestrian mall, the integrity of the smoke reservoirs must be maintained to prevent any smoke spilling out of the reservoir and flowing into the mall. This means that a fascia board must be provided to the same depth as the screens which form the smoke reservoirs. There would be some advantage in making this fascia board deeper than the other reservoir screens so that smoke spilling out of a reservoir would flow into adjacent reservoirs and not into the pedestrian malls.

As stated in Chapter 3, the requirements for the screens which form the smoke reservoirs are that they must be non-combustible, reasonably gas-tight and they must remain unaffected by intermittent exposure to flame. Small leaks in the screens, for example where pipes pass through, are not of great importance. The same requirement applies to a fascia board, but in this case even small leaks should be avoided.

(c) Smoke reservoirs alone will be of no long-term value in the control of smoke unless there is also smoke extraction or smoke venting from each. If the building is single storey, then roof venting is practicable. If it is multi-storey, natural venting will require the provision of vertical ducts through the upper storeys, of similar size* to that required for roof vents. This is unlikely to be economically acceptable, and mechanical extraction of the smoke may be possible and should be considered. In the case of both natural venting and mechanical extraction, the sizing of the system will depend on the design conditions selected, notably on the distance between the bottom of the smoke reservoir and the floor, but also to some extent on the depth of the smoke reservoir.

---

* Vertical ducts acting as chimneys can have smaller cross-sectional area than plain roof vents. The quantity of gas passed is proportional to the duct area multiplied by the square root of the height.

Natural venting

The following Tables give the design data for natural venting:
NB. These relate specifically to venting from shops, e.g. from the fire area. Different values apply to malls:

Table 4.1   *Areas of vents (in square metres) required in smoke reservoirs in shops*

(Depth of smoke reservoir) Height of vent outlet above lower edge of smoke reservoir screen (m)	*Area of vent required from each reservoir for various heights of smoke reservoir screens above floor (m²)*				
	2.5 m	*3 m*	*3.5 m*	*4 m*	*5 m*
1	6.0	8.1	—	—	—
1.5	5.0	6.6	—	—	—
2	4.4	5.8	7.3	8.9	—
3	3.6	4.7	5.9	7.2	10.1
5	2.8	3.6	4.6	5.6	7.8
8	2.2	2.9	3.6	4.4	6.2
10	1.9	2.6	3.2	3.9	5.5

It should be emphasized that the vent area decided by the above Table must be provided for each smoke reservoir and the vent area required does not depend on the area of the smoke reservoir. Even if the smoke reservoir has a small area it still requires the full vent size determined for the height.

The vent areas given in the above Table have been calculated on the assumption that the maximum fire size will be $3 \text{ m} \times 3 \text{ m}$ (i.e. perimeter 12 m), the size appropriate to a sprinklered occupancy. The vent area determined from Table 4.1 is a *total area* and this should be achieved by using several vents of smaller area to make up the total area, unless the smoke reservoir is deep. If a large vent opening is used with a shallow smoke reservoir, the action of the vent is not fully effective [16] because of the air entrained into the rising smoke plume at the centre of the vent; thus causing the vent to extract a mixture of smoke and fresh air instead of all smoke (Fig. 4.3). If the vent is kept small this entrainment of air is reduced and indeed can be eliminated. For this reason, if the smoke reservoir is shallow the total vent area should be made up of a number of small vents and not one large opening.

Table 4.2 sets out the minimum number of vents which should be used to make up this total area and from this Table it will be seen that the largest number of vents is required when the reservoir is shallow. The figures given in the Table assume that the vent will be evenly distributed over the smoke reservoir area.

Fig. 4.3 *Fresh air sucked into large outlet in a shallow smoke layer.*

Table 4.2 *Minimum number of vent outlets from a single smoke reservoir in a shop*

Distance between lower edge of screen and floor (m)	Distance between lower edge of screen and centre of vents (i.e. depth of smoke reservoir) (m)						
	1	1.5	2	2.5	3	3.5	4
2.5	5	2	1	1	1	1	1
3	6	2	1	1	1	1	1
3.5	8	3	2	1	1	1	1
4	9	4	2	1	1	1	1
5	10	5	2	2	1	1	1

If the natural exhaustion of smoke from a smoke reservoir is to be properly effective, provision must be made for fresh air to flow into the shop to replace the smoke and hot gases which are flowing out. It is essential that these fresh air inlets are low down in the shop so that the movement of the induced flow of air into the shop will not cause disturbance of the smoke/clear layer interface at the bottom of the smoke reservoir.

The area of the fresh air inlet openings should be at least twice the area of the outlet vents in any single smoke reservoir. Fresh air drawn in from the pedestrian malls will be satisfactory provided free flow is ensured, i.e. there is not an imperforate glass or board shop front with the possibility of closed doors.

It will be seen, on consideration, that in order to confine the smoke from a fire in a shop to that shop alone large areas of roof venting will be required, which even in single storey buildings, may be regarded as impractical. In multi-storey development the provision of space in the upper floors for ducts to vent smoke from the lower floors is likely to be difficult. Consequently it is suggested that the smoke extraction can be better achieved by mechanical means.

Mechanical extraction

For mechanical extraction of smoke from shops Table 4.3 gives the necessary design data. As with natural venting, these rates of extraction must be provided from each smoke reservoir. It may, however, be sufficient to devise a selective system which arranges for the full rate of extraction (see Table below) to be brought into operation in the smoke reservoir directly above the fire; with a reduced rate of extraction (say, half the values below) to be available for operation in adjacent smoke reservoirs. This would deal with any smoke spilling over from the reservoir above the fire, and avoid stagnant zones.

Table 4.3   *Rates of extraction required from smoke reservoirs in shops*

Height of bottom of smoke reservoir above floor (m)	Mass rate of smoke extraction required (kg/sec)	Maximum temperature of hot gases (above ambient) (°C)	Volume rate of smoke extraction required (at max. temp) (m³/sec)
2.5	9	560	22
3	12	420	24
3.5	15	340	26
4	18	280	29
5	25	200	35
6	35	150	44

The temperatures given in Table 4.3 are maximum values for the gases immediately above the fire. There is quite rapid cooling of these gases as they flow away from this point, partly by passing through sprinkler sprays, partly by radiation and convection to ceiling and floor, and partly by heat removed by extraction of the hot smoke and gas. Precise calculation of this cooling cannot be performed, because of uncertainties in the boundary conditions (e.g. heat loss coefficient, sprinkler cooling, etc.), but an approximate estimate [17] suggests that an extraction fan system capable of withstanding temperatures of 200–250°C should be sufficient provided the extract points of the system are well distributed in the smoke reservoir.

The required rate of extraction does not depend on the size of the smoke reservoir, but the values given in the Table are based on the assumption that the fire size will be limited by sprinklers (to a size 3 m × 3 m). The extraction points in the smoke reservoir should be as high as possible and preferably should be distributed evenly over the area of the reservoir. The values given in Table 4.2 should be taken as a guide to the number of extraction points to be provided. From this Table it will be seen that for deep smoke reservoir conditions a single extraction point is adequate but for shallow screen several extraction points must be provided.

As with the natural venting, arrangements must be made to allow fresh air to flow into the shop to replace the extracted smoke. The system adopted can either be by the provision of fresh air inlets, through which fresh air will be drawn into the shop, or by a fan-operated supply system to replace the extracted smoke.

If it is proposed to use air inlet openings then the rules which apply are identical to those for the natural venting, i.e. the air inlet openings must be low down, they must be double the area of the smoke extract vents which would have been required if powered extraction were not in use, they should be in the shop unless there is always direct connection at low level between the malls and the shops. However, there is scope for using smaller air inlet openings, provided the extract fan power is increased to overcome the extra resistance of the small inlets. This is an option which cannot be applied when smoke is removed by natural venting. In practice, the fan size required for smoke extraction is already large and any addition to the power for this purpose would probably be regarded as unwelcome.

When a fan-operated fresh air supply is proposed the mass rate of supply must be equal to the mass rate of extraction if the air supply is fed directly into the shop via supply grilles placed at low level. However, there may be some merit in considering the introduction of the fresh air into the malls by duct grilles. In this case there must always be direct connection at low level between shop and mall. If this is done, an air flow is set up from the mall into the shop which will act to prevent any stray smoke reservoirs in the shop from encroaching into the mall.

Full pressurization between shop and mall is not practicable, but the smaller air flow produced by feeding air into the malls and extracting from the shop on fire will produce an air flow pattern which will be a useful adjunct to the action of the smoke reservoir in the shop. If this smoke control method is proposed, the mass rate of air input to the malls should be greater (by a factor of about 2) than the mass extraction from the shop. The air supply in this case can be distributed along the mall (again at low level) so that air will always flow along the mall towards the shop on fire.

Having set out the conditions which are required if the smoke control proposal is to confine the smoke to the shop of origin, it must now be stated that although this is always to be regarded as the preferred system, nevertheless, the venting or extraction requirements are recognized as being, in some circumstances, onerous. This is particularly the case for small shops, while for large shops, i.e. those above 1000 m², the smoke condition dictates that the control system must confine the smoke to the shop of origin. The next section considers the alternative method of smoke control in which it is accepted that smoke from a fire in a shop will spread into the pedestrian malls.

4.3.2 Smoke extracted from the pedestrian malls

It is always an overriding principle that the pedestrian malls must be usable for a reasonably long period in the case of a fire in the shopping centre, because they form the connection between the shops and the open air and are, therefore, the main escape routes for the general public who will use the shops and who may be present in large numbers. If smoke is to be allowed to flow out of a shop on fire into the pedestrian malls, then the lateral spread must be limited, and the smoke must be kept at a high level above the pedestrians' heads so that they can move below it and be able to see where they are going.

In moving out of a shop into a mall, the smoke will have become cooler and will have mixed with more air during this movement. Consequently, the figures suggested for venting or extraction of smoke from a shop cannot be applied to a mall. The design requirements for smoke control in pedestrian malls are given in the following paragraphs (also see Fig. 4.4).

(*a*) Sprinklers must be installed in all the shops, and in the malls if there is any likelihood of display material or other combustibles being placed therein. This is essential because the smoke control design is based on a fire size of 3 m × 3 m.

(*b*) The ceiling of the pedestrian malls must be divided up into smoke reservoirs by placing screens across the mall and by providing each shop front with a fascia board of the same depth. The maximum length of the smoke reservoirs so formed should be 60 m (i.e. the screens must not be more than 60 m apart). The maximum area should be 1000 m² as in the case of those in shops, but this condition does not allow the 60 m length requirement to be relaxed. The depth of the smoke reservoirs should be as great as possible, subject to the provision that the bottom of the screens must be at least 2.5 m above the floor. The minimum depth for the smoke reservoir should be 1 m. Any space above a suspended ceiling may be included in the depth of the smoke reservoir provided it is perforated. The perforations should be at least 20% of the total area; if they are 40% or more then the whole of the reservoir can be above the false ceiling, but if they are less than

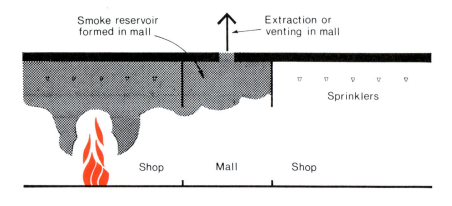

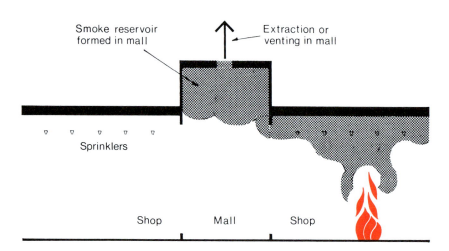

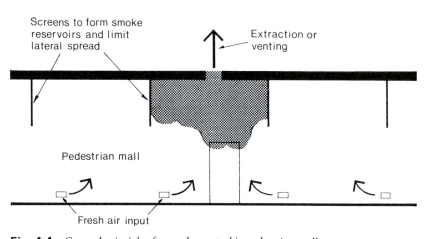

Fig. 4.4 *General principles for smoke control in pedestrian malls.*

40%, then the screens should project below the false ceiling so that some of the reservoir is below this level. For the space above to be included in the smoke reservoir the volume occupied by services, etc. should not be greater than 50% and no part of the service components should be made of a combustible material.

In some recently-designed developments, upstands spanning the full width of the mall have been used as an alternative to curtains to form the smoke reservoirs [10] (Fig. 4.5). In these cases the rules are similar to those given

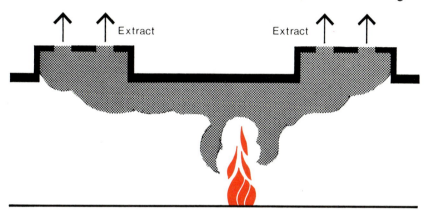

Fig. 4.5   *Upstands in mall used instead of ceiling screens to form smoke reservoirs.*

above. The distance between the centre lines of the upstand spaces should not exceed 60 m (200 ft). The depth of the smoke layer in the malls can be decided in exactly the same way as if curtains or screens were used to form the reservoirs. All the venting or extraction must be in the upstands which must be big enough to accommodate the vents designed as for reservoirs formed with screens (see below). The rules regarding false ceilings also apply as stated in the paragraphs above.

(c) The smoke reservoirs will not be effective in controlling the height of the smoke layer unless the smoke and hot gases are removed at the same rate as they enter. The removal can be by the natural movement of smoke through roof vents (particularly suitable for single storey malls) or by powered extraction.

Removal of smoke by vents

In this case the size of vents required will depend on two factors – the distance between the floor and the bottom of the screen which forms the smoke layer, and the distance between this point and the centre of the vent opening.

The vent areas required are given in Table 4.4.

Table 4.4   *Area of vents required for removal of smoke from smoke reservoirs in pedestrian malls*

Height of vent above lower edge of screen (i.e. depth of smoke reservoir) (m)	Area of vent for various heights of smoke reservoir screen above floor (height of bottom of smoke layer) (m²)					
	2.5 m	3 m	3.5 m	4 m	5 m	6 m
1	12	16				
1.5	10	13				
2	9	12	15	18		
3	7	9	12	14	20	26
5	6	7	9	11	16	21
8	4	6	7	9	12	16
10	4	5	6	8	11	15

The vent area decided from the above Table must be provided for each and every smoke reservoir, and the vent area does not depend on the area of the smoke reservoir, i.e. a smoke reservoir of small area requires the same area vent as one of the maximum size. As always the factor controlling the vent size is the assumed fire size of 3 m × 3 m.

The vent area given in the above Table is a *total area* and for shallow smoke reservoirs this should be made up of several small vents evenly distributed throughout the smoke reservoir. Table 4.5 indicates how many vents should be used to achieve this total area.

Table 4.5   *Minimum number of vents to use to make up total vent area for a single smoke reservoir in a pedestrian mall*

Height of lower edge of screen (distance from floor) (m)	Depth of layer of hot gases (m) Depth of smoke reservoir (m)						
	1	1.5	2	2.5	3	3.5	4
	Minimum number of outlets to make up total area						
2.5	9	4	2	1	1	1	1
3	12	4	2	2	1	1	1
3.5	15	6	3	2	1	1	1
4	18	7	3	2	2	1	1
5	24	9	4	3	2	2	1
6	34	12	6	4	3	2	1

Removal of smoke by powered extraction

The removal of smoke from the smoke reservoir by means of venting may not always be convenient, for many reasons associated with the planning of the shopping complex; and the use of mechanical extraction may often be the only practical solution. This is almost certain to be the case when the development is multi-level or multi-storey.

The volume rate of smoke extraction needed depends mainly on the distance between the bottom of the smoke reservoir screens and the floor (it being accepted, of course, that in all the design data given the fire size is assumed to be 3 m × 3 m). Table 4.6 below sets out the volume rate of extraction needed for the various depths of clear layer likely to be considered in the design of a mall smoke control system.

Table 4.6   *Rates of extraction required from smoke reservoirs in pedestrian malls*

Height of lower edge of screen of smoke reservoir above floor (m)	Mass rate of extract (kg/s)	Max. temp. of hot gases (above ambient) (°C)	Volume rate of extract (at Max. temp.) (m³/s)
2.5	18	280	29
3	24	210	34
3.5	30	170	39
4	36	140	44
5	50	100	56
6	70	75	73

The rates of extract given in the above Table relate to a single reservoir and this extraction capacity should be available to operate from any smoke reservoir affected by smoke or fire. It may be possible to operate a selective system so that the extract capacity of a smoke reservoir remote from the fire is brought into use at the smoke reservoir affected by fire or adjacent to that affected by fire.

As in the case of shops, if a selective extraction arrangement is used, then the full required extract capacity should be available from the smoke reservoir affected by fire; and the two adjacent smoke reservoirs may have a reduced rate of extraction (say, half the full value), so as to deal with any smoke spilling into them from the reservoir receiving smoke from the shop on fire.

It must again be emphasized that the rate of smoke extraction does not depend on the size of the smoke reservoir. The full rate, as specified in Table 4.6, must be used even if the smoke reservoir is much smaller than the maximum length of 60 m.

Air inlets

As in all smoke control systems, fresh air must be allowed to enter the mall at low level to replace the smoke which has been vented at high level, otherwise the venting arrangements will not be fully effective. The general rule is that the openings through which fresh air will be drawn should be at least twice the total area of the vent outlets from a single smoke reservoir.

There are several alternatives for fresh air entry:

(*a*) Doors at the ends of the malls are suitable provided it can be arranged that they open in the case of a fire, and fresh air flowing along the malls towards the fire may serve to prevent smoke flowing out of the active smoke reservoir. A disadvantage here is that part of the door opening is at high level, and air entering at this level may cause the smoke layer to be disturbed by mixing at the interface, so bringing smoke down to an unacceptable level in the mall. Additionally, this effect will be greatly increased by adverse wind action, but the use of ceiling screens spaced [10] several metres away from the door can act to minimize the effect (see Figs. 4.6 and 4.7, also Section 4.3.4).

(*b*) Fresh air inlets in the mall walls leading direct to open air (possibly by way of ducting) will be a satisfactory method, but planning considerations may make this impracticable.

(*c*) Windows or ventilators at the rear of shops will be another possibility, but the method adopted should not induce mixing between the layer of hot gas and the clear layer below, nor should it draw more smoke out of the shop on fire.

(*d*) The use of mechanically-operated fresh air supply may be the best way to overcome the disadvantages of the several possibilities mentioned above, but if this system is used the inlets should be as numerous as possible and as well distributed as possible, so that the fresh air flows into the malls at a low velocity.

(*e*) The suggestion that roof vents remote from the fire could be regarded as sufficient opening for fresh air input may seem attractive but it is not to be recommended. If smoke flows out of the immediately affected reservoir into adjoining reservoirs, it will be fairly cool and its buoyancy may not by sufficient to overcome the flow of fresh air in if no other inlet openings are provided. This will result in cool smoke being carried down to low level and possibly causing confusion in the malls.

Air input when mechanical extraction is used

When the removal of smoke from the smoke reservoirs is by powered extraction, the mass rate of removal of hot gases should be at least equal to the mass rate of supply of fresh air. When the air supply is also by a powered system, then the design rate of supply can be readily arranged. This has the

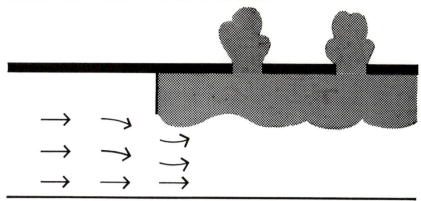

Fig. 4.6 *(a) flow of air along a mall reduces the possibility of smoke flowing beneath the screen.*

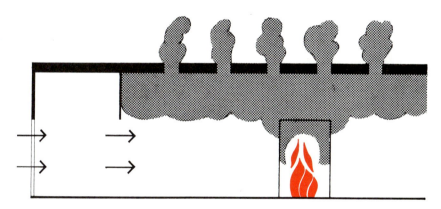

Fig. 4.6 *(b) Screen sited a few metres along mall to reduce wind effects.*

merit that it will be free from disturbance by external wind conditions and the general rule that air should be introduced at low level and at low velocity should, as far as possible, be observed.

When the fresh air supply proposed is by induced flow drawn in the specially arranged openings, then the rule which applies to the smoke removal by venting should be used. This stipulates that the area of opening for fresh air supply should be twice the area of the roof vents; consequently, an estimate of the roof vent area which would apply to the particular circumstances should be obtained and hence the area for fresh air inlet. It

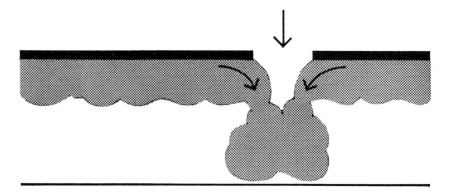

Fig. 4.7 *Low-level mixing induced by inflowing cold air: this situation must be avoided if possible.*

may be possible, in exceptional circumstances, to reduce the air inlet areas when powered extraction is being used but this is not recommended. The ready flow of fresh air along the pedestrian malls forms such a valuable and essential part of the smoke control system that no limitations on this air supply should be imposed.

4.3.3   Particular requirements in multi-level developments

In large shopping centre developments it is very probable that the design proposals will include several storeys of shops and pedestrian malls, and this feature can produce extra difficulties in smoke control.

1. A multi-storey shop opening on to two or more levels of pedestrian malls. If the smoke control system being proposed is that smoke will be confined to the shop of origin by venting or extraction from smoke reservoirs in the shop, the fact that extra levels are present will not impose any differences in design procedure. If within the shop there is fire-resisting separation between the levels, this will also mean smoke separation between the levels and the smoke reservoir, and smoke extraction systems will be designed separately for each level (Fig. 4.8a).

If there is no fire-resisting separation between the levels in the shop, then the smoke reservoirs should be formed at each level beneath the solid part of the floor and ceiling, and appropriate extraction provided for these (Fig. 4.8b). Where there is a void, or clear connection between two levels, then a smoke reservoir must be formed directly above this (or each of these) and the extraction from it properly sized taking account of the extra height between

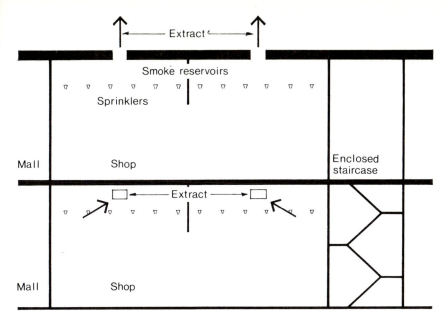

Fig. 4.8 *(a) Two-storey shop with separation between floors: smoke control as for a single storey shop.*

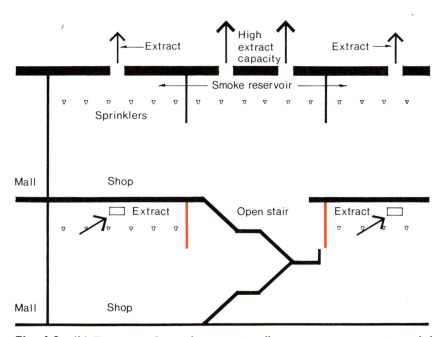

Fig. 4.8 *(b) Two-storey shop with open stair well: greater extract capacity needed above open well, unless screens (shown in red) are provided.*

the ground floor and the bottom of the smoke reservoir screen at the upper level. This reservoir will require a correspondingly greater level of smoke extraction (see Table 4.3) and the fresh air supply for the shop should take account of this greater extraction capacity needed.

In addition, the reservoir above the gap or void through which smoke can spread upwards should be both wider and longer than the horizontal opening of the gap in order to take account of the fact that a rising plume of smoke will spread sideways at an angle of approximately 15°. Since the size of any smoke reservoir is limited in length to 60 m or 1000 m² in area, the above requirement will set a maximum permissible size on the opening or void between two levels of a mall system. An alternative arrangement would be to provide a screen projecting downward from the lower level ceiling at the void edge so that smoke is prevented from travelling up the void. The depth of these screens should be at least 2 m and an extract system designed as described in Table 4.3 must be provided in the smoke reservoir(s) so formed in the lower floor [23].

It is suggested that the number of levels connected by open wells should always be limited to two.

2. Two or more levels of pedestrian malls with complete fire resisting separations between each level. In this situation the major part of the design of the smoke control system does not differ from a single storey development. Each level can be treated separately as a single storey complex and the smoke reservoirs, smoke removal and air input designed accordingly. The only difference is that the lower storey is unlikely to be vented for the smoke removal from the reservoirs. Whatever system is used for removing the smoke must be designed so that there is no possibility of the smoke discharged affecting either an upper level of shops and mall or any surrounding buildings.

3. Two or more levels of pedestrian malls with open voids connecting levels. This kind of development presents new difficulties relating to smoke control in the pedestrian malls. The problems arise because smoke flowing out of a shop into a lower mall, then rising in a plume to the upper levels will, because of the additional entrainment of air which occurs while it is moving along the lower mall and upwards, increase very considerably in volume. Some model experiments indicate that this increase can be up to tenfold, and unless the smoke control is properly designed, this vast quantity of smoke could spread sideways at the upper mall level or levels, and so cause smoke-logging in these (Fig. 4.9). The smoke control design must produce a smoke flow pattern as shown in Fig. 4.10; that is, the smoke flowing out of the lower mall up through the void should rise to the ceiling reservoir without spreading sideways into the upper levels.

Research work is still in progress on these problems and very precise

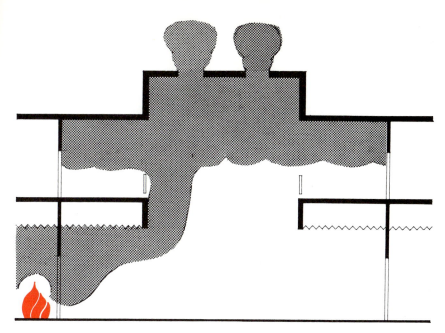

Fig. 4.9 *Smoke-logging of upper levels in a two-storey mall system.*

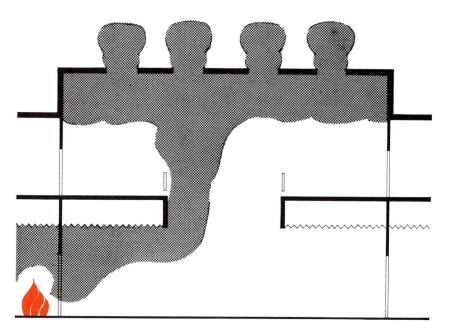

Fig. 4.10 *Good design prevents smoke spread into upper levels.*

design information is not yet available. It is, however, possible to set out a framework on which to base the design of smoke control systems for such situations.

(*a*) Smoke reservoirs must be formed in those parts of the pedestrian malls which have solid ceilings of single storey height. The design requirements for these reservoirs and the smoke extraction from them must be sized according to Tables already quoted for single storey malls.

(*b*) Where there is a void giving a direct opening connecting upper and lower pedestrian malls the following steps are suggested in order to ensure reasonable smoke control.

  (i) Limit the lateral size of the smoke plume which rises through the void to the upper storeys. This can be done by providing channels, using ceiling screens, along which the smoke will flow from the shop into the void. The spacing of these screens will be determined by the size of the shop fronts. (See Fig. 4.11.)

  (ii) Limit the vertical height through which the smoke can rise by arranging that no more than two storeys are linked by large open voids.

  (iii) Above each void form large smoke reservoirs which should be as deep as possible and wider and longer than the void below. The smoke from any shop should then be channelled to the void edge by arranging suitably spaced screens, and the volume of smoke which must be vented or extracted can be estimated from the curves [7] in Fig. 4.12 in conjunction with Table 4.6. If this graph does not cover the required range, an estimate should be made by extrapolation, if necessary, by referring to Reference [7].

(*c*) An alternative method is to use smoke curtains or screens hanging below the edges of the connecting voids (at each mall ceiling level) to confine the smoke to a reservoir on the level of origin and to extract smoke from that reservoir. This method would make it possible to reduce the overall extract capacity and would make feasible malls of more than two levels. The depth of these smoke curtains, or screens, must be at least 2 m and the required extract capacity is calculated as already described for single storey malls [23].

(*d*) It is essential that a plentiful supply of replacement fresh air be provided on both levels. The rules for assessing the amount of fresh air required or the area of inlet for replacement air should be those suggested for a single storey mall with an addition of at least 50% in volume supply or inlet area. In making the calculation, the smoke reservoir over the two-level void should be considered and if the same air supply serves more than one such void, then the void with the largest vent area or extraction rate must be taken. The fresh air inlets must be divided between the two levels, more air being introduced into the upper level than the lower level. It is suggested that

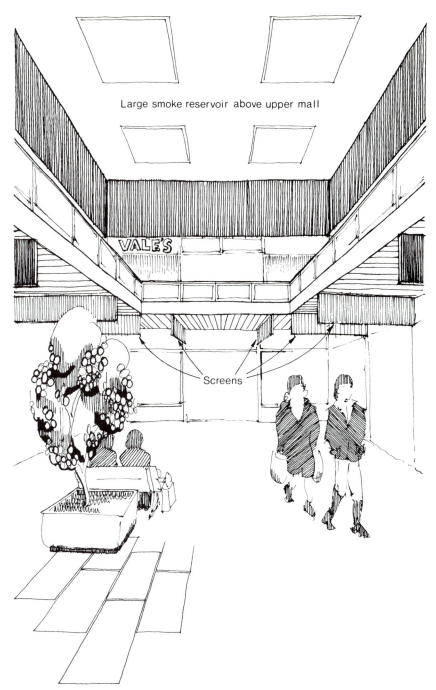

Fig. 4.11 *Illustration of screens in ground floor shopping mall.*

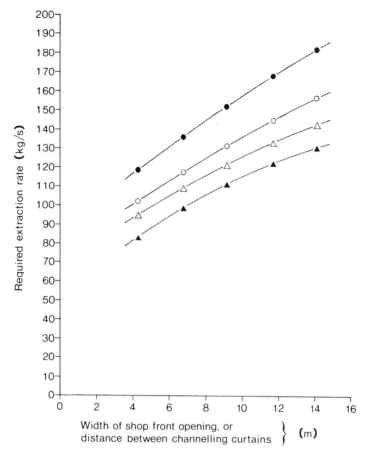

Fig. 4.12 *Required mall extraction rates for various shop front widths (14 m internal width). (Fire size $Q = 5\ MW$; ambient temperature $T_1 = 15°\ C$).*

$\bullet X_r = 4.0\ m$ $\triangle X_r = 2.5\ m$
$\bigcirc X_r = 3.0\ m$ $\blacktriangle X_r = 2.0\ m$

where X_r = height from upper floor to base of smoke layer.
Note : 1. The upper floor is taken to be 5 m above the lower floor.
2. The required depth of channelling curtains would be $\geqslant 1.5$ m for this case.

the ratio of replacement air on the upper level to that on the lower level should be 3:2.

The rule for the introduction of fresh air at a low level should be followed in each of the two interconnected malls, and if the source of air supply is by automatically-opening doors or shutters, then it is suggested that these be opened immediately on both floors wherever the fire occurs even if initially the smoke does not spread into the two-level void.

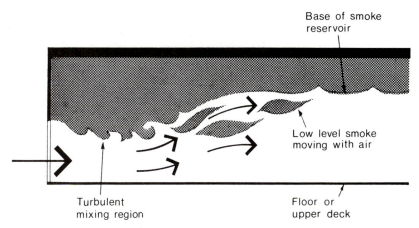

Fig. 4.13   *A door acting as an air inlet.*

4.3.4   Further design points

The interaction between an air-inlet (e.g. a door) and a smoke reservoir [18]
A moving airstream in contact with a region of stationary air (or smoke, or gas) will have a lower pressure than the stationary air – this is the Venturi effect. Hence, a moving airstream will attract the stationary air towards itself. The force of attraction increases with increasing velocity of the airstream. In Fig. 4.13 the clean airstream flowing through the door attracts the smoke towards itself. The smoke reservoir base follows the surface of constant pressure, and bulges downwards near the door. Unless the reservoir base is high enough above the door (at least $1\frac{1}{2}$ m for reasonable safety in a multi-storey mall) the smoke base will come just below the top of the door. Turbulence at the airstream/smoke interface will cause smoke to mix into the incoming air. This smoke will then move with the air, travelling through the mall at low level.

For a constant air volume-flow-rate, increasing the airstream cross-section will reduce the velocity. This will in turn reduce the forces attracting the smoke towards the airstream. Fig. 4.14 shows a smoke-restraining curtain set back from the door. The curtain's bottom edge is higher than the door. The incoming airstream will increase in vertical section as it travels from the door to beneath the curtain, and so will slow down. There is less tendency to pull down the smoke-base. There is less turbulence at the smoke/air interface mixing smoke into the airstream and any such mixed smoke is at a higher level.

Using a horizontal shelf (Fig. 4.15) instead of a curtain permits the same

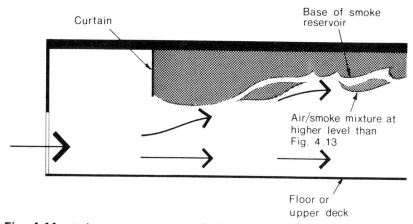

Fig. 4.14 *A door acting as an air inlet but with a smoke-restraining curtain set back from it.*

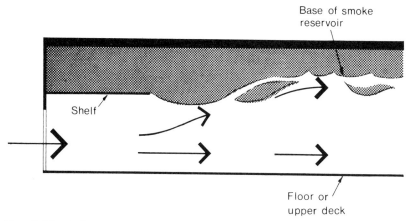

Fig. 4.15 *A door acting as an air inlet but with a shelf moving the air/smoke interface back from it.*

air expansion and gives the same results. In either case the height of the shelf (or the height of the lower edge of the curtain) should be more than 3 m above the floor; the shelf-edge (or curtain) should be more than 3 m back from the door. (These distances are tentative and can be relaxed if necessary.)

A further advantage in both cases is the reduction of the disturbing effects that external wind gusts have on the smoke reservoir, especially on the turbulent mixing at the smoke/air interface.

Wind effects on smoke venting

Many existing and proposed shopping complexes rely on natural venting for their smoke extraction. The discharge of smoke from the vents can be affected by wind, in particular by the pressure patterns which can be set up in the vicinity of tall buildings. Reference to this problem is made in Chapter 3, and the suggestions made there apply equally to this Chapter.

The main suggestions are:

(*a*) if possible vents should discharge vertically with, if necessary, suitable weather protection;

(*b*) the planning of outlets near the base of tall buildings, or buildings extending to a higher level than the discharge point, should be avoided;

(*c*) vents should not be placed in vertical, exposed, outward-facing walls, but walls opening into large internal wells or similar spaces, protected from the wind, could be used.

Investigations [19] are in progress into the wind pressure patterns caused by tall buildings in conjunction with flat roofs, but the number of possible building shapes is so large that it is unlikely that any general rules other than those given above will emerge. An adverse wind is not the only air disturbance feature which can affect a smoke control system. Internal factors within the building must sometimes be considered, expecially where they contain items which can move large quantities of air in a random way.

An underground railway system would be an example, where the normal operation involves a pneumatic effect of large quantities of air being moved in both directions. For example, if an enclosed shopping complex, with a smoke control system, had an open direct link through an underground station concourse to the tube system below, then the movement of the trains would cause a random and massive air movement effect on such a smoke control system and could completely destroy its designed function (Fig. 4.16). In such cases there must be separation between the area affected by such a massive air movement and the shopping centre's smoke control system. It should also be appreciated that this kind of air disturbance will affect mechanically-operated smoke control systems as well as those relying on natural ventilation.

Air curtains

It has been suggested to the authors that air curtains might be used to prevent smoke travelling from one part of a shopping complex to another. There is no evidence to suggest that air curtains can be used in this way and, in fact, their very action is likely to bring the high-level smoke down to the floor of the mall or shop. It is therefore strongly suggested that when air curtains are used in shops or malls in which smoke control by the methods described in this

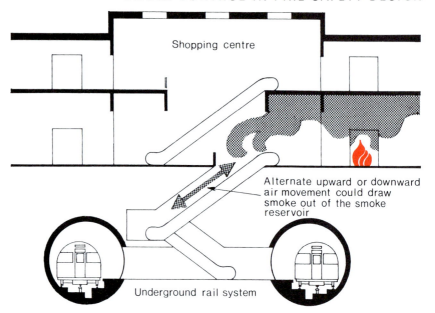

Fig. 4.16 *Example of uncontrolled air movement disturbing the smoke control system.*

Chapter is installed, then the air curtains must be stopped when the emergency smoke control system is brought into operation.

Flow of smoke beneath a mall (or corridor) ceiling

In the consideration of smoke control in situations described in this Chapter it is sometimes necessary to estimate the rate at which a layer of smoke will spread sideways when there is no extraction or venting of the smoke. The three-dimensional problem of the radial spread of smoke which occurs when a plume of smoke and hot gases from a fire meets a ceiling is difficult to resolve, but the simpler two-dimensional case of smoke flowing beneath a corridor ceiling has received study and the results may be of some importance in the design of smoke control systems.

Hinkley [2] has given a tentative theory for the flow of hot gases along a corridor (or mall). He makes the assumption that the smoke layer will consist mainly of air heated by entrainment into the rising hot gases over the fire, and that any effects on the flow due to differences in composition between the gases in the layer and the cool air below can be neglected. He considers the case where the hot gases are flowing only in one direction along the mall. In practice, the hot gases will generally be flowing in both directions away from the fire and it may be assumed that, when there is no draught, half the total flow of gases will be in each direction.

The velocity of the gases in the layer will be determined by the buoyancy, frictional and inertial forces, the transfer of momentum to the cool air by mixing, and finally, viscous forces. Hinkley suggests that the viscous forces may be neglected and that the flow pattern will then be a steady state problem and determined by the Froude number, which is the ratio between the inertial and buoyancy forces, since to a large extent it may be assumed that the frictional forces and the transfer of momentum by mixing are also determined in this way. Developing his theory in this way, Hinkley produced an expression for the mean velocity of the gas layer moving under a ceiling in a mall. This expression is:

$$\bar{v} = 0.8 \left(\frac{gQT_c}{C_p P_a T_a^2 W} \right)^{1/3} \tag{4.1}$$

where $\bar{v}$ = mean velocity of the gas layer
g = acceleration due to gravity = 981 cm/sec^2
Q = heat output of the fire
T_c = absolute temperature of the gases beneath the ceiling
T_a = absolute temperature of the air in lower part of mall
C_p = specific heat of air at constant pressure = 1000 J/Kg°C
P_a = density of the cool air = 1.2 Kg/m^3 at 20°C
W = width of the corridor
and   0.8 = an empirical constant.

It is pointed out that this expression is for the mean velocity of the layer; the velocity of the leading edge will be lower than this and will decrease as the distance from the fire increases, because of the cooling of the gases. Initially, the layer will flow beneath the ceiling without mixing with the air beneath, although some entrainment of smoke into the cold air near the nose of the advancing layer may occur. As the velocity of the nose decreases due to cooling, the depth of the layer will increase to maintain the rate of flow of gases. The following conclusions are drawn from this treatment of the problem.

(*a*) The rate of movement of the smoke layer depends only slightly on the rate of entrainment of air by the fire (i.e. on the perimeter of the fire), but is roughly proportional to the cube root of the heat output of the fire, and is inversely proportional to the cube root of the width of the mall.

(*b*) the depth of the smoke layer in a mall of given height is roughly proportional to the perimeter of the fire, and inversely proportional to the product of the velocity of the layer and the width of the mall.

(*c*) When a fire has grown beyond a critical size, for which the layer of hot gases occupies half the height of the mall, the flow becomes unstable near the fire and smoke-logging to ground level may occur.

(d) Even when the fire is smaller than this critical size, there may be some mixing of hot smoky gases from the nose of the advancing layer into the flow of cold air which is moving towards the fire; thus the layer of smoke will deepen.

This treatment of the problem assumes that flow is occurring along a corridor or mall in which there are no irregularities or obstacles. Morgan [20] has extended the theory to take account of obstacles in the corridor and has derived the expression.

$$\bar{v} = 0.96 \left(\frac{gQT_c}{K^1 C_p P_a T_a^2 W} \right) \tag{4.2}$$

in which the factor K^1, described as a velocity pressure coefficient, depends on the obstacles in the mall. For the case of the smoke flow when the end of the corridor is a door smaller in area than the corridor, Morgan shows that the value of K^1 will be given by:

$$K^1 = \frac{d_1}{C_v d_s} \left(\left[\frac{W d_1}{w C_v d_s} \right]^2 - 1 \right) \tag{4.3}$$

where C_v is the coefficient of discharge and all the other quantities are as

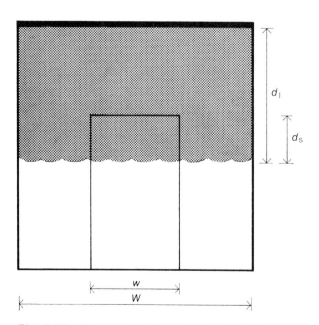

Fig. 4.17 *Diagrams showing symbols used in Equations on page 104.*

shown in Fig 4.17. When the door is the same height as the mall the expression for K^1 reduces to:

$$K^1 = \left(\left(\frac{W}{w}\right)^2 - 1\right) \tag{4.4}$$

For fairly gentle obstacles to flow (e.g. bends in the corridor) the values of K can be deduced for the values which are known for pipe flows [21] which should be applicable to corridors.

REFERENCES

1. Home Office. (1972). Fire precautions in town centre developments. Fire Prevention Guide No. 1, London, HMSO.
2. Hinkley, P. L. (1970). The flow of hot gases along an enclosed shopping mall. A tentative theory. Fire Research Note No. 807, Fire Research Station, Borehamwood, England.
3. Hinkley, P. L. (1971). Some notes on the control of smoke in enclosed shopping centres. Fire Research Note No. 875, Fire Research Station, Borehamwood, England.
4. Phillips, A. M. (1971). Smoke travel in shopping malls. Model studies Part 1. Rates of lateral spread. Fire Research Note No. 864, Fire Research Station, Borehamwood, England.
5. Heselden, A. J. M., Wraight, H. G. H. and Watts, P. R. (1972). Fire problems in pedestrian precincts. Part 2. Large scale experiments. Fire Research Note No. 964, Fire Research Station, Borehamwood, England.
6. Wraight, H. G. H. (1973). Fire problems in pedestrian precincts. Part 4. Experiments with a glazed shop front. Fire Research Note No. 997, Fire Research Station, Borehamwood, England.
7. Morgan, H. P., and Marshall, N. R. (1975). Smoke hazards in covered multi-level shopping malls. Part 1. An experimentally based theory for smoke production. Building Research Establishment Current Paper 48/75, Fire Research Station, Borehamwood, England.
8. Thomas, P. H., Hinkley, P. L., Theobald, C. R. and Simms, D. L. (1963). Investigations into the flow of hot gases in roof venting. Fire Research Technical Paper No. 7, London, HMSO.
9. Morgan, H. P., Marshall, N. R. and Goldstone, Mrs B. M. (1976). Smoke hazards in covered multi-level shopping malls. Some studies using a model 2-storey mall. Building Research Establishment Current Paper 45/76, Fire Research Station, Borehamwood, England.
10. Hinkley, P. L. (1975). Work by the Fire Research Station on the control of smoke in covered shopping centres. Building Research Establishment Current Paper 83/75, Fire Research Station, Borehamwood, England.
11. B. R. E. Digest 173 (1975). Smoke control in single-storey shopping malls. Building Research Establishment, Fire Research Station, Borehamwood, England.

12. National Fire Protection Association (1959). *Operation school burning.* Official report on tests conducted by the Los Angeles Fire Department.
13. D'Auria, C. (1973). Protection against fire in garages. Anticendio e Protezione Civile 9/73. 640–642. Fire Research Station Library Translation No. 328 (1974), Borehamwood, England.
14. Bullen, M. L. (1974). Effect of a sprinkler on the stability of a smoke layer beneath a ceiling. Fire Research Note No. 1016, Fire Research Station, Borehamwood, England.
15. Morgan, H. P. (1977). Heat transfer from a buoyant smoke layer beneath a ceiling to a sprinkler spray. Fire Research Note No. 1069. Fire Research Station, Borehamwood, England.
16. Spratt, D. and Heselden, A. J. M. (1974). Efficient extraction of smoke from a thin layer under a ceiling. Fire Research Note. No. 1001. Fire Research Station, Borehamwood, England.
17. Heselden A. J. M. (1977). Private communication.
18. Heselden, A. J. M. (1977). Private communication.
19. B. R. E. Annual Report (1975). Page 39. Building Research Establishment, Garston, England.
20. Morgan, H. P. (1977). The flow of buoyant fire gases beneath corridor ceilings. Fire Research Note No. 1076, Fire Research Station, Borehamwood, England.
21. I. H. V. E. Guide (1970). Vol. C. Published by The Chartered Institution of Building Services (previously the Institution of Heating and Ventilating Engineers), London.
22. *Fire Prevention.* No. 93. December 1971.
23. Morgan, H.P. and Marshall, N. R. (1978). Smoke hazards in covered, multi-level shopping malls: a method of extracting smoke from each level separately. Building Research Establishment Current Paper 19/78. Fire Research Station, Borehamwood, England.

5   Smoke control on protected escape routes – pressurization

5.1   GENERAL

In normal fire prevention design the intention will always be to confine the fire within a fire-resisting compartment, and smoke control measures for this part of the building have already been discussed (Chapter 3). In making the escape, the occupant will move towards a door which leads to the protected part of the escape route, generally a corridor, lobby or staircase. The enclosing construction of this part of the escape route will be fire-resisting, and the door leading to it from the fire area will also be fire-resisting. Thus the name – protected part of escape route – implies it has full structural protection from the effects of the fire to the standard prescribed by the relevant regulations.

Within this protected part of the escape route there will be no restrictions on the distance over which the occupant has to travel to reach the final place of safety and indeed the route may well be needed for use by people from parts of the building not involved in the fire. It is clear, therefore, that this part of the escape route must remain usable for a relatively long period during the course of the fire and that for this reason no smoke, or at least the very minimum amount of smoke, must encroach into this enclosure. However, as stated, the separating element between the fire and the protected route is a door, which has cracks around its edge large enough to allow the ingress of smoke, and which in performing its function will be opened from time to time. Unless some sort of smoke control or ventilation is provided for the protected route it will inevitably become contaminated by smoke and it is therefore considered essential that smoke should be prevented from moving into this escape route, or that it should be so diluted as to create no visibility or toxicity hazard. A visibility of at least 5 m would normally be regarded as essential.

Regulations and Codes make some recommendations for smoke control measures in staircases, lobbies and in some cases corridors, all of which may form the protected escape route. The requirements using natural means of ventilation are quite specific; those for mechanical means of smoke control have, until now, been unspecified, but the publication of a new Code of Practice [1] gives very precise design guidance.

5.2 NATURAL VENTILATION

When it is proposed to use natural ventilation to control the movement of smoke on escape routes, reliance is placed on climatic conditions to create the necessary air movement.

The various factors which cause smoke from a fire to move about in a building are:

(a) the smoke's own buoyancy;

(b) the air movement in a building caused by weather conditions (i.e. wind and temperature differences inside and outside the building); and

(c) the air movement caused by any mechanical ventilation or air-conditioning system.

The first of these, smoke buoyancy, will almost certainly (see Fig. 2.6, Chapter 2) cause smoke and hot gases to move into the escape route (i.e. the staircases, lobbies and corridors), although if these are some distance from the fire and the smoke has had time to cool, this movement may be sluggish.

The second factor, weather conditions, is the only agency available to counter or minimize the effect of this smoke ingress for buildings with naturally ventilated escape routes. This factor is unpredictable and uncontrollable. Estimates are available which show that for an appreciable part of the time, up to 25%, the wind-induced air movement will be too low to move adequate fresh air into the building via the permanent or openable windows, so that the smoke is diluted or removed to a satisfactory extent. An investigation [2] carried out at the Fire Research Station has shown that for smoke control purposes, ventilation in one wall of a space (staircase or lobby) is unlikely to be effective even when wind velocities are quite high and the conclusion reached is that cross-ventilation (i.e. openings in opposite walls of the space to be ventilated) is essential if smoke control is to be achieved by natural means. Even so, a reasonable wind velocity in the right direction is required and this is not available for a significant proportion of the total time.

The use of manually openable windows as a means of smoke control on escape routes in buildings also has its critics as it is by no means certain when or by whom they will be opened. The escaping occupants may not be able to reach them because of smoke or they may be too occupied to think of doing so, and it will be much too late for escape purposes if they have to wait for the Fire Brigade to do the opening. It follows that for smoke control by natural ventilation to have any chance of success it must consist of cross-ventilation using permanent openings and some of the Codes and Regulations do ask for this. However, such an arrangement can well create uncomfortable conditions for the occupants and the effect can be negated by the unofficial covering of the ventilation openings by the occupants who may not realize that these openings are provided for their safety. It will also conflict with the conservation of energy and the requirements for thermal control. The danger of the unauthorized closure of permanent ventilation openings is well recognized, and the risk can be reduced by the provision of windows or shutters which are normally closed but which automatically open when smoke is detected at the door leading to the escape route. This device can take the form of a drop sash or of a horizontally pivoted window; both will be held closed by a solenoid type catch but will immediately fall or swing into the open position if smoke is detected, or if a fault (such as current failure) develops in the operating system. An example of such a window is shown in Fig. 5.1 [3].

5.3 MECHANICAL VENTILATION

When the provision of effective means of smoke control by natural ventilation becomes difficult, then the use of mechanical ventilation for this purpose must be considered. If it is necessary to maintain the escape route free of smoke, the use of a mechanically driven extraction system to remove any smoke that may have entered the escape route is not satisfactory, because the action of the extraction system will reduce the pressure level in the escape route (stair, lobby or corridor). This will draw more smoke into that space, thus tending to reduce visibility, rather than improve it. (See Fig. 5.2.)

The only satisfactory system is one which will completely prevent the smoke from entering the escape route, and this can be done by ensuring that an adequate flow of fresh air is set up which will drive the smoke and hot gases away from the escape route. To do this, fresh air is injected into the staircase or lobby (or both); it then leaks out of these spaces via door cracks, moves across the fire area and escapes from the building by window cracks, or by a specially provided opening. This is the system which is now usually called *pressurization*, because in order to set up the required air flow, the pressure in the escape routes is raised above that in the rest of the building. The Regulations and Codes already referred to give this method as a possible alternative to natural ventilation and a Code of Practice [1] relating to its use has recently been published.

Fig. 5.1 (a) *Building in which automatically openable windows have been used as an alternative to permanent ventilation.*

Fig. 5.1 (b) *Close view of automatically openable window used, showing permanent ventilation on floor above.*

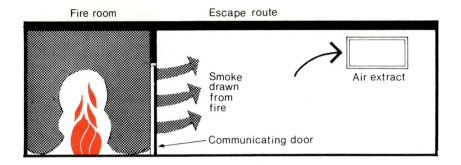

Fig. 5.2 *Mechanical ventilation and smoke movement: (a) extraction only: smoke drawn into escape route.*

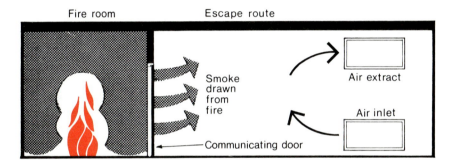

Fig. 5.2 *(b) extraction and air supply: smoke still drawn into escape route.*

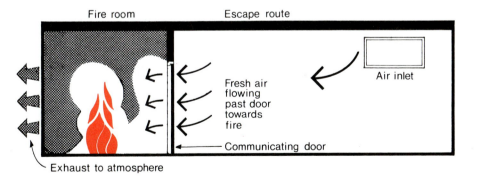

Fig. 5.2 *(c) air supply only: no smoke in escape route.*

5.4 BASIC WORK AND TESTS FOR VALIDATION OF PRESSURIZATION

The idea of using a system of pressurization in a building in order to establish a predictable air flow pattern is not particularly new. It was used during World War II in certain important buildings to ensure that in the event of enemy action involving poison gas or bacteriological sprays, these agents would not infiltrate into the vital control rooms of the defence organization. In peace time it has been used to obtain a dust-free atmosphere in work rooms; and in buildings where radioactive gas or dust could escape, it can ensure that the area of contamination is confined. In hospitals it is regularly used to ensure sterile atmospheric conditions in operating theatres.

As a result of these applications there was by the late 1950's a considerable amount of experience available on the design of pressurized spaces in buildings and in 1961, Hall (P.S.A. Fire Department) suggested that these techniques might be used to control smoke movement in the case of fire. A year or two later he was able to incorporate his ideas in a government building then being designed. At about the same time the same ideas were being suggested in Australia and when, in 1957, the Height of Buildings Act was amended so that, in the main, the height limit was removed from buildings [4], a Code of Practice was produced which allowed three possible methods of smoke control for stairways. These were: (1) balcony approach; (2) lobby approach; or (3) pressurization. This, as far as is known, is the earliest mention of pressurization in a Code of Practice relating to fire precautions.

These two uses of pressurization in case of fire in buildings pre-dated any of the research on the use of pressurization in fires by several years although, at the time they were suggested, information about the pressures developed in fires was becoming available. There is no doubt, however, that they both played a considerable part in awakening interest, almost worldwide, in pressurization as a fire precaution and research on general aspects of the suggestions was undertaken in various countries during the latter part of the 1960's.

5.4.1 Tests carried out in the U.K.

In a department store

In 1964, the Fire Research Station carried out a preliminary investigation into the feasibility of using pressurization to control the movement of smoke in buildings [5]. A newly built 3-storey department store was placed at their disposal for this purpose by Marks & Spencer Ltd. This store had an enclosed staircase which served all three floors. Doors opened from the stair at each floor and communicated with the accommodation. A single fan (rated at 3000 ft^3/min (1400 l/s), placed at the head of the stair on the second floor was used for pressurization. A small room on the ground and first floors was constructed to serve as the 'smoke room' in the four tests carried out.

The smoke was generated by a specially designed apparatus in which controlled combustion of known types and quantities of cellulosic materials produced smoke of the type experienced in fires. The report on these tests states that these generators do not produce smoke in a quantity or at a temperature normally associated with fires, but they do provide a consistent source of warm smoke, forming a useful tool for experiments and simulating smoke conditions in the early stages of a fire. Four tests were carried out using the warm smoke, and in two tests the staircase was pressurized; in the other two no ventilating fans were used. For each of these conditions, smoke was generated first on the ground floor and then on the first floor. In the two tests which had no fans operating, smoke leaked past the door between the stair and smoke chamber so that even with the limited amount of smoke being produced, conditions in parts of the staircase became unbearable in some 10 to 20 minutes. When the door between the smoke chamber and stair was opened the inrush of smoke was considerable and the stair and landing were quickly filled with it, to the extent that smoke now penetrated another closed door and spread on to other parts of the accommodation. It is also reported that opening a vent at the top of the stairs did not have any significant smoke clearance effect, which is of general interest as such a requirement is included in most building legislation.

For the two tests in which pressurization was applied to the staircase enclosure, the degree of pressurization was varied, so that some idea could be obtained of the minimum pressure differential which would stop the passage of smoke through the door cracks. It was found that an excess pressure of 0.02 in w.g. (5.0 Pa) was sufficient to prevent the penetration of smoke through the door cracks. It was also found that with this value of excess pressure, smoke did not spread onto the stair when a door was opened briefly, as would be the case if a person passed through it.

It was concluded from these experiments that an excess pressure of 0.05 in w.g. (12.5 Pa) in the escape routes was sufficient to keep these spaces clear of smoke, and that the opening of doors (other than that leading to the fire) for short periods did not appear to have any significant effect on the overall effectiveness of the pressurization system. This level of pressurization did not completely prevent the flow of smoke into the staircase when the door between the smoke room and the staircase was open. The tests also demonstrated the smoke clearance action of pressurization, in that smoke allowed to enter the escape route by opening a door was cleared in a relatively short time when the door was closed again.

At the Fire Research Station

Following the preliminary investigation described above, the Fire Research Station at Borehamwood carried out a series of tests on the effects of pressurization, using a 4-storey experimental building (Fig 5.3) which was available for fire tests [6]. This building had a staircase which led into accommodation at each floor level. In these tests the staircase, i.e. the escape route, could be pressurized by the use of two fans connected to a duct system which allowed fresh air to be introduced into the staircase at each floor level. The fans could be used individually or both at the same time. The source of smoke used in this work was a wood crib fire in one of the ground floor rooms adjacent to the staircase. The fire load density used was 6.2 lb/ft² (30 kg/m²) and the window opening was such that the fire load per unit ventilation opening was 46 lb/ft² (220 kg/m²). This is the fire load likely to be found in domestic occupancies, offices, hotels and so on.

The experimental work carried out had several parts. First the pressure developed by a fire was measured. The height chosen for this measurement was that corresponding to the top of a normal door, i.e. 6 ft 6 in (2 m) from the floor. As would be expected, this pressure depends on the temperature developed in the fire room; for the experimental set-up used, the pressure developed by a fire reached a limiting value of about 0.025 in w.g. (6 Pa). However, the report also explains that factors other than fire temperature could vary this pressure. The position of the neutral plane is one factor; it can be changed by altering the position and size of openings into the room. For instance, a large gap at the bottom of the door can lower the neutral plane and so increase the pressure at the top of the door.

The second part of the work described was a study of the pressure differentials which can be produced by adverse weather conditions in a building, and a study of these was made in the six winter months during which the whole experimental programme was carried out. It was found that the maximum pressure differential (caused by weather) obtained between the smoke chamber and the stair was 0.05 in w.g. (12.5 Pa). This value is similar to that reported by other workers, who found that in a ten-storey block with a 20 m.p.h. (32 km/h) wind the pressure differential across a door between a room and corridor was of the order of 0.01–0.02 in w.g. (2.5–5.0 Pa). Consequently, it was realized that the pressure differentials due to weather conditions could be higher than those developed by a fire, and in the light of the measurements made it was decided that pressure differentials of the order of 0.1–0.2 in w.g. (25–50 Pa) would easily override the differentials reported above.

The third part of the work consisted of measurements of the air flow necessary to put into a staircase or other enclosure in order to produce pressure differentials of this order across the doors leading from it. It is assumed that the door is the common 'penny fit' in its frame, i.e. the crack

Fig. 5.3 *Building used in fire test of pressurization at Borehamwood.*

round it is $\frac{1}{8}$ in (3 mm) so that the area of crack round a normal single leaf door will be 20 in² (130 cm²) and both measurements and calculations show that an air flow of 160 ft³/min (75 1/s) past the door will produce a pressure differential of 0.2 in w.g. (50 Pa). This indicated that the air supply needed to pressurize an escape route was relatively modest, and that more work was needed to obtain detailed information about the air leakage past various building components.

The fourth part of the programme investigated the effectiveness of pressurization to control smoke. Tests using smoke produced by a generator were used in the initial stages of the work but the conclusive tests were those in which the smoke produced by wood crib fires was used. In these tests, with the door between the fire and stair closed and when no pressurization was applied to the stair, this space became completely smoke-logged and unusable in 11 minutes. In 18 minutes, flame had penetrated into the staircase and in 25 minutes the half-hour fire check door had completely collapsed. When the test was repeated with precisely the same structural conditions, i.e. a half-hour fire check door, and with the same fire, but with pressurization applied to the stair to a level of 0.2 in w.g. (50 Pa), then no smoke penetrated on to the stair for the whole of the 50 minutes for which the test was continued. No appreciable temperature rise was recorded on the staircase during this period and although the edge of the door burnt away so that the crack size was in fact increasing during the test, there was no fire penetration to the staircase during this 50 minutes.

The fifth and last part of this work was reported in *Fire Research* (1969) [7] and is concerned with the effect of opening the door between the fire and the pressurized space. A fire test similar to that used in the above work was staged, and the conditions which obtained in the staircase enclosure when the door to the fire is opened to various extents was compared with and without pressurization being applied to the stair to a level of 0.2 in w.g. (50 Pa), measured with the fire door shut. Photographs taken during this test show clearly that pressurization is still effective in controlling smoke movement on to the stair even when the door between the stair and the fire is open.

An important additional conclusion which can be drawn from this work is that not only does pressurization constitute a satisfactory and practicable method of smoke control, but pressurization also improves the fire resistance performance of the door between the fire room and the staircase [6].

5.4.2 Theoretical studies of smoke movement

An experimental survey of the behaviour of smoke in a large variety of existing and occupied buildings, with many possible types of fire is clearly not practicable. Therefore, a theoretical survey of any given practical situation is an attractive proposition. The availability of computer techniques makes this a possibility, but the main difficulty is specifying the precise temperatures, air flows and pressure gradients for each type of fire. Several workers [8, 9, 10, 11, 12 and 13] have written computer programs which appear to satisfy the observed behaviour in the few instrumented fire tests which have been reported, but validation of these simulations for large complicated buildings which have a pressurization system still needs to be done. Even so, the availability of the computer programs as a design tool is a valuable contribution to the overall understanding of the behaviour and limitations of a pressurization system.

5.4.3. The practicality of pressurization

Following the fire tests carried out at the Fire Research Station, the Heating and Ventilating Research Association (now the Building Services Research and Information Association) at Bracknell carried out an investigation into the problems associated with installing pressurization in buildings.

This investigation looked at the following problems:

1. The pressures acting on and in buildings which can influence the internal air flow patterns and hence the efficiency of pressurization.
2. The influence of mechanical ventilation installations other than pressurization systems, on internal air flow patterns.
3. The air leakage characteristics of building construction so that air supply requirements for pressurization could be determined.
4. The requirements of air distribution plant for pressurization to achieve the design objectives.
5. The reliability of pressurization systems.

To study items 1 and 2 above, theoretical procedures using computer techniques were used. For item 3, information was collected from published data and in addition, site tests to measure air leakage characteristics in buildings were carried out. Items 4 and 5 consisted of a study of the technical and economic requirements for distributing air to pressurized escape routes. The results of this work, which still ranks as one of the most comprehensive and complete surveys, are fully described in a Report by P. J. Hobson and L. J. Stewart [14], but the scope of the main findings is set out below.

The report studies the four main influences which provide motive force for air movement within buildings – wind, stack effect, mechanical ventilation and fire. It concludes that the pressure developed due to weather, stack and fire can be counteracted by pressure developed by mechanical ventilation systems, and that the orders of magnitude concerned are similar to those reported in the earlier work [6, 15 and 16]. Considerable details are given as to the leakage rates found for the several building components concerned, e.g. windows, doors and lift shafts and numerical values are suggested for design purposes. The importance of allowing the pressurizing air to escape is emphasized, and design rules for several methods are given. The equipment, fans, ducts and terminals, is discussed; the features important to pressurization are outlined. All of the results reported are used as the basis of the design requirements laid down in the Code of Practice just published; these are all given in detail later in this Chapter.

5.4.4 Research overseas

Australia

Although New South Wales was probably the first in the world to incorporate specific proposals for pressurization of escape routes in Codes of Practice [4], there do not appear to be any published records of research having preceded this action. A recent bulletin from the Commonwealth Experimental Building Station [16] briefly described work done in connection with 'Flame Spread and Smoke Tests'. This involved some limited tests in a three-storey experimental tower and it reported that in the conditions of air flow and pressure required in the N.S.W. Codes, serious smoke pollution of the pressurized area will be prevented. They further reported differences in effectiveness of a pressurization system according to whether the doors were open or closed.

Canada

There has been an extensive programme of research into the problems of smoke control in tall buildings carried out by the National Research Council of Canada, in Ottawa. The work has been both experimental and theoretical [17–24]. Concentrating on the air movements produced in buildings by weather conditions, they deduced from this study the effect both of mechanical pressurization and natural ventilation. The culmination of these studies was an explanatory document [25] setting out the principles of smoke control and the requirements for seven alternative methods of achieving them in practice. These range from natural ventilation to fully pressurized buildings, but pressurization in some form or other appears in five of the suggestions.

The paper recommends that smoke control measures should be included in all buildings over six storeys high (with some exceptions) and the proposals do not permit a sprinkler installation to be an alternative to a smoke control system. The basis for the smoke control measures is to maintain tenable conditions in those parts of a building where occupants may have to remain during a fire emergency and in exit stairs and certain elevators. Tenable areas have been defined as those where no more than 1% of the atmosphere has originated from the fire area. The suggestions given in this paper have now been incorporated in Canada's Regulations for smoke control measures in buildings.

U.S.A

Little real research into pressurization has been done in the U.S.A., other than the fire tests reported below, but some requirements for pressurization are incorporated into the Unified Building Code (U.S.A.) [26]. These requirements are based on a system of smoke control developed by the Los Angeles Fire department [27, 28] which used both pressurization and extraction. This system has been criticized in several quarters [29].

Europe

The smoke control systems which have received attention in the investigations carried out in Europe have relied on a so-called 'scavenging' system in which a large air input and an equal air extract is applied to a lobby [30, 31 and 32], but simple pressurization systems have not been studied other than in the U.K.

5.4.5   Fire Tests

A major criticism of all the investigatory work into the use of pressurization for smoke control has been the absence of real fire experience other than those contrived on a laboratory scale. For this reason the results of three full scale fire tests have been studied carefully by all of those concerned with smoke control and they are described in some detail below.

In the summer of 1972, fire tests [33] were carried out in a 22-storey office building by the Polytechnic Institute of Brooklyn working on behalf of the New York City Fire Department. The 22-storey office building (Fig. 5.4) became due for demolition and the opportunity was taken to stage fire tests in it with a view to testing pressurization and smoke exhaust schemes. For the purpose of these tests, two fans were installed, a 40 000 ft³/min vane axial type blower was placed at the bottom of the stair shaft, selected for the test and a 10 000 ft³/min vane axial fan was installed at the top of the stair shaft. The large fan at the bottom was for pressurization, while the smaller fan at the top was for smoke exhaust. With this fan installation, pressure differentials of 1.0 in w.g. (250 Pa) were obtained at the bottom of the stair shaft, but with only a single supply point at the foot of the stair there was a pressure gradient up the stair, so that at the top of the shaft the pressure differential was only 0.3 in w.g. (75 Pa). When the pressure at the bottom of the stairs was reduced to 0.3 in w.g. (75 Pa), then the pressure at the top was 0.08 in w.g. (20 Pa).

All of the above figures apply to the condition in which all doors on the stair well were closed. When one or more of the doors on to the stairs was open, the pressure differential at levels above the open door was much

Fig. 5.4 *Twenty-two storey building used in fire test of pressurization in New York.*

reduced. A test in which the stair pressurization was provided by the fan at the top of the stair shaft with the bottom of the stair open, indicated that smoke (from a smoke candle) would concentrate at the foot of the stair. This was regarded as undesirable and, therefore, no further tests were done using this condition. Smoke tests, using cool smoke, were staged with pressurization supply air being injected at a low level, in order to establish that the pressurization system was effective in keeping the stairs clear of smoke. This it did satisfactorily.

Following this preliminary work, four fire tests were staged. One fire was arranged on the 7th floor and used a fire load of wooden desks, chairs, paper, rubber foam cushions and polyurethane scraps giving a fire load density of 6.3 lb/ft² in a test room of size 54 ft × 32 ft (1728 ft²). The other three fires were arranged on the 10th floor, each differently situated in relation to the stairwell. These used smaller fire areas (200–350 ft²) and had office furniture arranged to give fire load densities of 5.1, 9.0 and 5.1 lb/ft².

The conclusions reached in these tests are stated as follows:

(*a*) The feasibility of using stair pressurization as a means of ensuring smoke-free conditions in high-rise office buildings was demonstrated.

(*b*) Measurements of smoke levels and direct visual observations made in the stair indicated that with as many as three doors wide open the entire stairwell remained free of smoke.

(*c*) While the corridor and adjacent lobby areas on the fire floor were observed to have heavy smoke levels, the test stair provided a clear and safe passage for the evacuation of occupants, and an effective route by which the fire fighters could approach the fire location.

(*d*) Measurements made indicated that the elimination of the exhaust fan at the top of the stair would, with suitably sized exhaust openings, improve the performance of the pressurization system.

At about the same time, in July 1972, a somewhat similar test [34] was being arranged in Atlanta, Georgia by the City of Atlanta Building Department. Here the 14-storey Henry Grady Hotel became scheduled for demolition, and it was decided to stage fire tests in order to evaluate a pressurization system.

Fans were installed to pressurize the stair shaft and the elevator shaft. These were arranged to supply air to the foot of each shaft, 22 000 ft^3/min being the maximum feed possible for the stairwell and 37 000 ft^3/min to the elevator shaft, which was a common shaft housing 3 lifts. In addition, fans were installed so that an approach lobby (vestibule) to the stairwell on the fire floor could be either pressurized or maintained at a reduced pressure by an exhaust fan on the roof. With these sizes of fan, the stairwell pressure could be raised to 0.8 in w.g. (200 Pa) at the bottom, reducing to 0.1 in w.g. (25 Pa) at the top with all doors closed. An interesting feature of the pressurization pattern in the stairwell was that when the air flow was reduced to give a pressure of 0.2 in w.g. (50 Pa) at the bottom, the pressure at the top was still 0.1 in w.g. (25 Pa). In the elevator shaft with the fan running at its maximum, the pressure across the closed lift door on the fifth floor, the fire floor, was 0.05 in w.g. (12.5 Pa).

Smoke tests and fire tests were staged to examine the behaviour of the system for a variety of conditions. Three fire tests were staged, one on the fifth floor and two on the third floor. In one fire test the fuel consisted of 550 lb of miscellaneous items of furniture, while in the other two, wood pallets were used as fuel. In all of the fires the fire load density was of the order of 4.0 lb/ft^2. The level of pressurization used in the fire tests was 0.15 in w.g. (37.5 Pa) between the stairwell and the fire floor (this fell to 0.05 in w.g. (12.5 Pa) when the door to the stair was opened on the fire floor), and 0.05 in w.g. (12.5 Pa) between the elevator and lift lobby on the fire floor.

The conclusions derived from these tests are stated as follows:

(*a*) Exit stairway smoke protection by mechanical pressurization was satisfactorily accomplished.

(*b*) The effect of the vestibule protection was beneficial in providing additional compartmentation and resistance to smoke migration.

(*c*) The pressurization of elevator shafts is feasible and was effective in preventing smoke infiltration into the shafts.

(*d*) It was noted that excessive door resistance can be expected on exit doors that are located near a large fan which provides a single supply source for the pressurizing aim.

(*e*) Additionally, when this single supply point is low down in the stairwell opening the ground floor exit door gives an almost complete loss of the pressurization in the stairway.

(*f*) In view of these two points it is recommended that air be introduced at several, not one, strategic locations throughout the shaft. This will promote uniform pressurization and will eliminate significant loss of pressurization due to an open door at any point.

These two fire tests have served to fill a gap in the experience available concerning the value of pressurization for smoke control, and in this respect they are invaluable. However, they were staged in old buildings not built or designed for pressurization, and the pressurization system used in both cases was not installed in the building in the way to be expected of a system designed from the start as a working part of the building.

This criticism does not apply to the third fire test, now described. This was staged in a specially designed office building in Hamburg [35 and 36], which has six floors above the ground floor, and a basement (Fig. 5.5). In plan the building is roughly rectangular, measuring 28 m (92 ft) by 13 m (43 ft). On each of its two long sides it is attached to adjoining buildings, and at the rear it faces directly on to a canal (the Monkdammfleet). Pedestrian access to the building is, therefore, only possible on the other short side (the South side) of the building.

This particular set of boundary conditions posed great problems for the siting of the staircase and the services core with two lifts, because access to an external wall was not practicable and the only way considered possible to provide smoke-free stairs using traditional methods of natural ventilation was by using a 'smoke tower' (or large open well) in the middle of the building, so that access to the stair on each floor was via an open balcony in this open well. This method combined the uncertainties of smoke control relying on atmospheric conditions with an extravagant waste of floor area for the service core of the building; and a pressurization scheme was designed

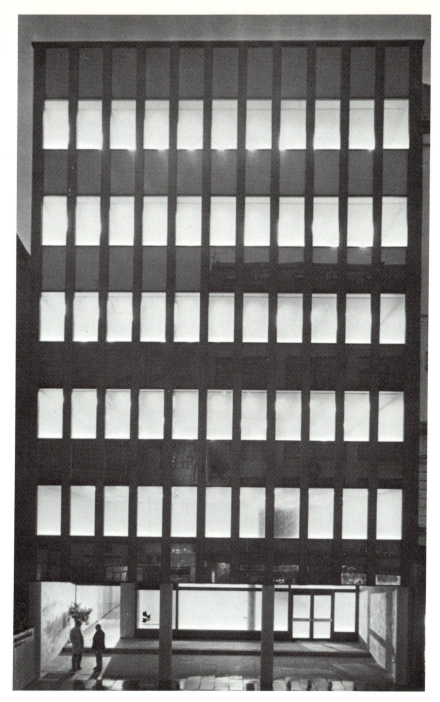

Fig. 5.5 *New building in Hamburg in which the pressurization system was subjected to a fire test.*

for the building, which made considerable economies in space and provided a smoke control system which could be regarded as completely reliable. (See Fig. 5.6.)

The Hamburg Building Control Officer and the Hamburg Fire Brigade had no experience of the use of pressurization as a means of keeping smoke away from escape routes, but they were prepared to consider new ideas, and when they were shown the volume of experience that had built up in other countries they finally agreed to grant building approval subject to these two conditions:

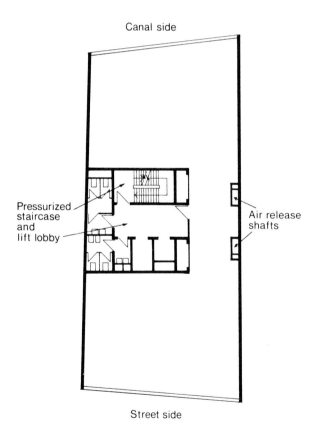

Fig. 5.6 *A typical floor plan of the building, showing the pressurization system.*

(*a*) That the pressurization scheme design must allow for four open doors without loss of pressure in excess of that provided by the fire or by external 'stack effect'. The design calculations submitted must demonstrate that the scheme would be effective under this condition.

(*b*) An actual fire test must be staged in the completed building to show that the pressurization installation was effective in keeping smoke away from the lobbies and staircase.

The pressurization system proposed was two-stage – that is to say, it incorporated a low level of pressurization in the lift lobby and in the staircase for continuous running for the whole of the time the building was occupied, and it also had an 'emergency state' in which the pressurization went to a higher level (in both the staircase and lift lobbies) in the event of a fire anywhere in the building.

The design of the system included the following features:

(*a*) It was a two-stage system, staircase and lift lobbies at the same pressure: 15 Pa for normal running, increased to 50 Pa for the emergency condition (Fig. 5.7).

(*b*) The emergency condition was initiated by smoke detectors, two of which were placed on each floor in the office accommodation area.

(*c*) Emergency electrical supply was provided by a diesel generator set housed in the basement.

(*d*) The staircase and lift lobbies were supplied from separate fans, each situated on the roof, with the air intake for each also on the roof.

(*e*) When an office-to-lobby door and/or staircase-to-lobby door was open, an air flow into the fire area from the pressurized area had a velocity of at least 0.5 m/s through the open door (Fig. 5.8). This condition was to hold when up to four doors were open, one of those four being the door to open air on the ground floor. In order to satisfy this condition the air flow into the staircase and into each lobby was increased, and an additional permanent leakage opening introduced so that the pressure in the staircase when all doors were closed, did not greatly exceed 50 pa.

(*f*) The release of the pressurizing air from the office accommodation on each floor was achieved by means of two vertical ducts shown in the plan of Fig. 5.6. These ducts were common to all floors, and a fire-resisting steel damper flap, normally closed, opened when the smoke detector on that floor opened.

(*g*) Thus, the pressurizing air was vented on the fire floor only. This localized venting reacted back on all the pressurization levels in the building and these were then calculated to be about 58 Pa on the staircase and 60 Pa in all lobbies on unvented floors. This increase in pressure levels was considered to be acceptable.

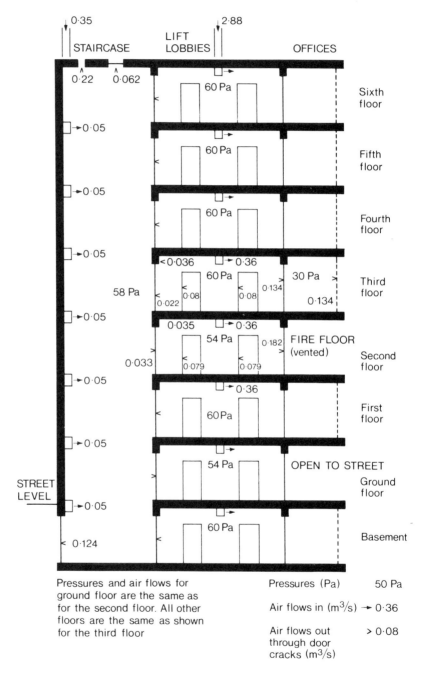

0·35

2·88

STAIRCASE

LIFT
LOBBIES

OFFICES

0·22   0·062

60 Pa   Sixth
             floor

0·05

60 Pa   Fifth
             floor

0·05

60 Pa   Fourth
             floor

0·05

<0·036   0·36

58 Pa   60 Pa   0·134   30 Pa   Third
                                    floor
0·022   0·08   0·08   0·134

0·05

0·035   0·36

54 Pa   0·182   FIRE FLOOR
0·033   (vented)   Second
0·079   0·079   floor

0·05

0·36

60 Pa   First
             floor

0·05

54 Pa   OPEN TO STREET
STREET   Ground
LEVEL   floor

0·05

60 Pa   Basement

0·124

Pressures and air flows for
ground floor are the same as
for the second floor. All other
floors are the same as shown
for the third floor

Pressures (Pa)   50 Pa

Air flows in (m³/s) → 0·36

Air flows out   > 0·08
through door
cracks (m³/s)

Fig. 5.7 *Emergency pressurization levels.*
126

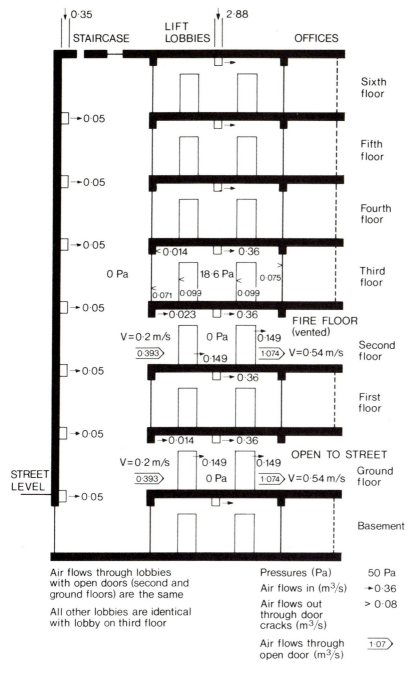

Fig. 5.8 *The calculated 'open door' air flow.*

Three types of test were carried out on the pressurization system in this building:

(*a*) A measurement of pressure differentials and air flows established in the building when the leakage areas which were expected to obtain in the completed building had been set up on each floor.

(*b*) A visual demonstration of the behaviour of the system using cold smoke generated in the fire room.

(*c*) An actual fire test of the building in which a full scale fire was staged in the office accommodation on the second floor.

The fire test was carried out under conditions which completely simulated the behaviour of the finished building; where windows or doors had not been fitted, temporary partitions were erected. These were carefully sealed and holes were made which were carefully sized to represent the estimated leakage of the component. The installation of the actual pressurization system was complete except for the duct terminals in the various positions. The fire room was set up on the second floor and was formed by partitioning off part of the accommodation on the north side of this floor. The size of the area so partitioned was some 4 m × 15 m. The lobby door opened into the fire area on this floor and the two vertical vent ducts were also included. Although the building will be sealed when it is completed, there will be, nevertheless, leakage away from the floor via the normal air conditioning ductwork and this leakage was simulated in the fire room partitioning by suitably sized leakage holes. The fire load used consisted of 370 kg ($\frac{1}{3}$ ton) of wood arranged in two groups of eight cribs with large slabs of expanded polystyrene between the cribs to provide dense black smoke.

The measurement of conditions in the building was very comprehensive:

(*a*) Pressure differential measurements were made between lobbies and staircase and lobbies and accommodation on all floors.

(*b*) Air flow measurements were made of the air supplied to each pressurized space by the pressurizing fans.

(*c*) Air flow measurements were made of the air passing out of each open door.

(*d*) Temperatures were recorded at several places in the fire room, in the lift lobby on the fire floor and at several points in the staircase.

(*e*) Smoke density measurements were made in the lift lobby, on the fire floor and at several points in the staircase.

(*f*) Concentrations of CO and CO_2 were measured in the lift lobby on the fire floor and in the staircase at the second floor level.

Measurements of air flow and pressure differentials were made on the day before the fire test and these showed that: Air flows supplied to the pressurized spaces (staircases and lift lobbies) were a little greater than the design values (by about 8%). Pressure differentials between the pressurized spaces and the accommodation spaces were greater than those used in the design calculations by about 40%. These high values were attributed to the very good fit of the doors between the lift lobby and office spaces and between the lobbies and staircase. These components were door sets with doors fitted into steel frames and the clearances round the edge of the door were small.

Cold smoke tests

Air flow velocities past the open doors were lower than those used in the design calculations by about 20%, with the exception of the door to open air on the ground floor; in this case the air flow out of this door was greater than calculated by about 100%. This was probably due to relatively small differences in the air flow resistance at the various doors which would have a large effect on the air flows. However, in spite of these differences it was considered that the air flows through the open doors would be adequate to give successful smoke control. The cold smoke tests gave a visual demonstration of the movement of the air out of the pressurized spaces and showed how the massive air flow through the open doors could provide a smoke-free space on the fire side of the door leading to the escape route (in this case the door to the lift lobby). At no time during any of the cold smoke tests was any smoke observed in the lift lobbies or the staircase even with four doors open (office-to-lobby, lobby-to-stair doors on second floor and stair-to-lobby door and lobby-to-open air doors on the ground floor).

A log of the observations made during the fire test is given in Table 5.1. It will be seen from this that temperatures of 700°C were recorded in the fire room about 15 minutes after the start of the test, which lasted in all about 36 minutes. During the whole of the test period the lobbies and staircase were recorded as smoke-free, except for a very short period (about 2 minutes) at the peak of the fire when smoke obscuration of 30% was recorded in the lift lobby on the second floor. This was, however, a transient condition and the lobby quickly cleared and remained clear of smoke for the rest of the test. The flaps into the vertical shaft were opened one minute after the fire test started and, at this stage, it was observed that air was coming down one shaft and feeding the fire. However, as the fire developed the reverse action took over and large amounts of smoke went out of the building by this path.

Pressure measurements taken at various points during the fire test showed that the stack effect of these shafts was appreciable, and pressures of the order of -20 Pa were developed in the fire room; because of this, the pressure differential recorded between the lift lobby and the fire room was about 90 Pa during the fire. This also affected the air flow when doors were open. When four doors were open (two on the fire floor and two on the ground floor) the air velocity through the open stairway lobby door was twice that

Table 5.1 *Record of fire test held at Grosser Burstah 46–48 on 4 April 1976*

Time (min)	Temperature over fire	Door condition	Smoke spread
0	Fire started	All closed	
1	1 pallet		
2	ignited		
3			
4	200°C		Smoke produced
5		Lobby/office door – 2nd	Fire-room smoke-logged
6	300°C	Floor open	
7		Lobby/stair door – 2nd	Slight smoke into lobby
8	500°C	Floor open	Lobby free of smoke
9			
10	400°C	Stair/lobby door open –	
11		Ground floor.	
12	400°C	Door to open air –	No smoke in lobby/stairs
13		Ground floor open – 4 doors open.	No smoke in lobby/stairs
14	2nd pallet		No smoke in lobby/stairs
15	now burning	Office/lobby doors on	Smoke moving into lobby
16		1st & 3rd floors open	30% obscuration in lobby
17	700°C		Stair free of smoke
18		2 other stair doors open	
19		(8 doors open) 1st & 3rd floors.	Lobby clear of smoke
20	630°C	Ground floor doors shut	
21			
22			
23	630°C		
24	630°C	Stair door 4th & 5th	No smoke in lobby or
25		floor open (8 doors	stair
26		open).	
27		Office doors 4th & 5th	No smoke in lobby or
28	620°C	floor open (10 doors	stair
29		open).	
30	510°C	Ground floor doors open	No smoke in stairs or
31		again (12 doors open).	lobbies
32			
33			
34			
35	Fire extinguished		

The fire was staged on the second floor. The fire load was 370 kg of wood arranged as small cribs with some polystyrene to give smoke. With all doors closed, the pressures in lobbies and stairs were a little higher than the design figures, but with doors open the air flows were a little lower. In the early stages of the fire air was being drawn into the fire room via one of the vertical shafts. The window in the fire room did not break and remained in position for the whole test. The flaps in the vertical shaft were opened on the *fire floor only*.

recorded in the air flow measurements on the previous day, and it was observed that air was flowing into the building at the ground floor door (compared with out of the building during the previous cold test). The window in the fire room remained intact during the whole of the fire test.

It is noteworthy that not only was the installation completely successful in keeping the escape routes clear of smoke, but that the fire itself, because of the venting by vertical shafts, developed an air flow which markedly assisted the pressurization system. The fire load crib arrangement was designed to achieve a temperature of 1000°C above it for at least 15 minutes, and two preliminary fire tests of the arranged fire load had shown that this could be obtained. However, because of the arrangement to exhaust the pressurizing air by the vertical duct system, and because there was a very generous supply of fresh air available to cool the fire, the design temperature of 1000°C was never reached. The maximum temperature above the fire was only just 700°C and this was only maintained for a minute or two.

The combination of the fire and the vertical exhaust ducts produced a negative pressure of 20 Pa in the fire room and this acted to draw fresh air in for cooling as well as materially assisting the pressurization system. The building and the pressurization system were designed as an integral whole and formed a very nice example of a comprehensive design for fire safety, showing that if such a system is introduced, fire resistance requirements can be reduced. For instance, the doors to the stairs and lobbies were both 90 min fire resistant and they played no part in keeping the fire and smoke away from the stairs, being open for most of the test.

5.4.6 Experience now available

Pressurization systems have now been installed in a large number of buildings both in the U.K. and overseas. The tall building with a central service core is an example of a design that has become more safely designed by the inclusion of mechanical ventilation (i.e. pressurization) for smoke control on the escape routes. The Pearl Insurance Building [37] in Cardiff shown in Fig. 5.9 is an example of one such building, and other buildings can be cited where it has been an economic advantage to avoid placing staircases and lobbies on the external walls.

It is clear that there is now, as a result of the accumulated experience drawn from research from fire tests and from the design of installations for buildings, a considerable fund of information about the use of pressurization in buildings as a positive means of smoke control. In consequence it is possible to set out in detail the steps which must be followed and the requirements which must be achieved in order to design a satisfactory system of smoke control by pressurization.

Fig. 5.9 *Building used by Cardiff fire brigade in pressurization tests to limit smoke movement.*

5.5   REQUIREMENTS FOR A PRESSURIZATION SYSTEM

The general intention of pressurization is to provide a graduated increase in pressure from rooms, through corridors and lobbies to stairs, which would give greater freedom from smoke to persons passing through each space towards the stairs. The pressurized space, which will usually be a staircase, in some cases with its associated lobbies, is protected in the event of fire, if:

(a) the integrity of the structure of the pressurized space is maintained;

(b) the integrity of the pressurization system is maintained;

(c) the flow is totally outward through the doors and other operational openings;

(d) air can be vented to outdoors from unpressurized spaces.

The installation of a pressurization system will usually be primarily for this purpose (i.e. smoke control) but it can also be used to meet certain ventilation requirements, e.g. the normal ventilation of pressurized spaces. For this, the pressurization system can be run at a low level and will run continuously or whenever the building is occupied. This system will be coupled to a high level of pressurization which will come into operation in an emergency only.

A separate pressurization system should be provided for each staircase, although a single plant and distribution system can be arranged to serve a staircase and associated lobbies when both are pressurized. The purpose of this is to avoid events (such as open doors) on one staircase interfering with the pressurization level on another. The required capacity of the plant for a pressurization system is determined from the requirement that the air flow from the pressurized spaces shall be totally outwards and continuous. This condition is met if the pressure differential across doors and other operational openings separating pressurized and unpressurized spaces is positive even when adverse conditions are set up by wind and fire effects.

There are three separate parts to the design of a pressurization system which are complementary to each other and completely dependent on each other. These are:

1. *The air supply to the pressurized space(s).* This must be a mechanically operated air supply, usually distributed by ductwork to the required places within the pressurized spaces. This air must be drawn from outside the building in such a way that contamination by smoke is not possible. It may not always be necessary for this air supply to be filtered, silenced or warmed.

2. *The air leakage out of the pressurized space.* The space being pressurized will unavoidably, in any building, have air leakage paths in its enclosing surfaces. These leakage paths will be the cracks round doors, cracks round windows, direct leakage through the building fabric, leakage through air conditioning ductwork and so on. If a pressure differential is maintained

between a pressurized space and its adjacent space(s), air will flow through these leakage paths. The volume rate of flow of air through these leakage paths will determine the level of pressurization which is maintained in this pressurized space.

3. *The venting of the unpressurized part of the building.* It is absolutely essential that the air flowing out of the pressurized space to the unpressurized part of the building shall then be able to leak out of the building to open air, either by leakage of the building itself or by means of specially contrived vents. This leakage path must present a low resistance to air flow so that only a small pressure differential is developed across it. If this leakage path is not provided, then it will not be possible to develop a pressure differential between the pressurized space and the rest of the building.

5.5.1 Design sequence for a pressurization smoke control system

If the three elements described above are provided, they must bear the correct relationship to one another if a satisfactory smoke control system is to be established. The first step in the design is to decide which spaces, stairs, lobbies, corridors, are to be pressurized.

Spaces to be pressurized

1. *Staircase(s) only.* The simplest scheme will be to pressurize the staircase (or staircases) only. This, although simple, will give limited smoke control and in general should only be used when the horizontal part of the escape route on every floor is relatively short. It must be understood that the smoke control will be limited to the vertical part of the escape route. In general, it should only be used when the staircase is approached on each floor directly from the accommodation or through a small, simple lobby. This lobby should not give access to lifts or to rooms (e.g. toilets) which could constitute an appreciable leakage path and allow pressurizing air to bypass its required direction of flow.

The effect of an open door on the pressurization will be reduced if the staircase is approached by a small, simple lobby. This lobby should be unventilated and it will, in consequence, be pressurized by the air flowing out of the staircase. Figs. 5.10 and 5.11 show two examples of pressurizing the staircase only and show how the protection is limited to the staircase.

2. *Staircase and all or part of the horizontal route.* In every building in which each floor has a horizontal component of the protected route (other than the small simple lobby mentioned in (1) above) the pressurization should be carried into the lobby, and possibly into the corridor beyond. In this case all the spaces concerned (staircases, lobbies and/or corridors) will have their

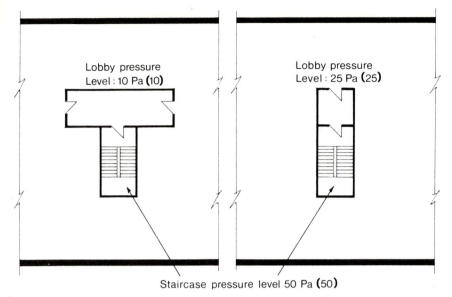

(a) Pressurization levels with the lobby/accommodation doors shut.

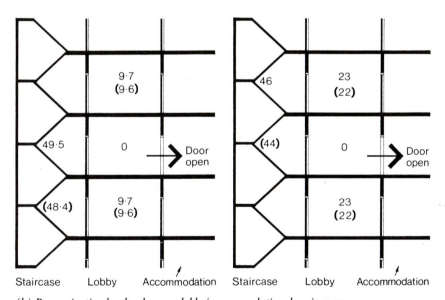

(b) Pressurization levels when one lobby/accommodation door is open.

Fig. 5.10 *Pressurizing staircase only. For a ten-storey building, pressure levels (Pa) are shown unbracketed. For a five-storey building, pressure levels (Pa) are shown in brackets.*

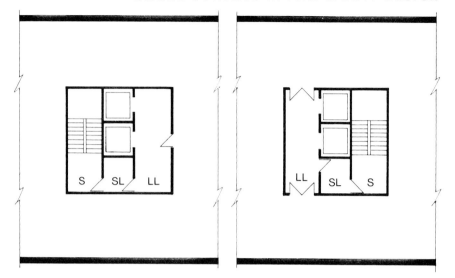

Fig. 5.11 *Staircase only pressurized – gives no protection to lift lobby. Pressure levels developed in lobbies by staircase pressurization (Pa):*

Building height (storeys)	5	10	Building height (storeys)	5	10
Staircase (S)	50	50	Staircase (S)	50	50
Stair lobby (SL)	25.6	26.2	Stair lobby (SL)	25.1	25.2
Lift lobby (LL)	1.2	2.7	Lift lobby (LL)	0.2	0.4

own mechanically driven supply of fresh air. By this means the protection is carried right up to the door leading into the accommodation area in which a fire might occur. Additionally, and most importantly, the effect of an open door on the pressurization levels is largely confined to the floor concerned. Fig. 5.12 shows an example of a pressurization system involving staircase and lift lobby on each floor and the effect of an open door on one floor is shown.

3. *Pressurizing lobbies and/or corridors only.* In some buildings it may be found necessary, perhaps for constructional reasons (such as difficulty in arranging the required ductwork for independent pressurization) to allow the staircase to be pressurized by the air which leaks onto it from the associated pressurized lobbies or corridors. If properly designed, this can be a satisfactory method; but in some cases it may be found that the total air supply needed for pressurizing the lobbies only may be greater than that required if the staircase and lobbies are independently pressurized. If the staircase is not pressurized (except by air leakage from lobbies) it should not be permanently ventilated except by reason of any opening which may be

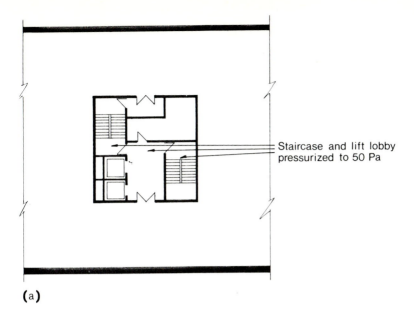

Staircase and lift lobby
pressurized to 50 Pa

(a)

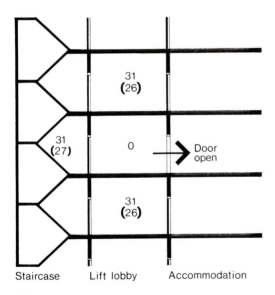

31
(26)

31
(27)

0

Door
open

31
(26)

Staircase   Lift lobby   Accommodation

(b)

Fig. 5.12 *Pressurizing staircase and lift lobby: pressure levels with lift lobby/accommodation door open shown in (b). For a ten-storey building, pressure levels (Pa) are shown unbracketed. For a five-storey building, pressure levels (Pa) are shown in brackets.*

found necessary in the design calculations to satisfy the open door condition (which is described later).

Operation of the pressurization

Having decided on the spaces in a building which are to be pressurized the next step is to determine how the system is to operate and what pressurization levels are to be used.

1. *Single-stage or two-stage.* The pressurization system can be designed to operate only in an emergency (e.g. in case of fire). At all normal times there will be no excess pressure developed in any of the spaces chosen for pressurization. This is called a *single-stage system.*

Alternatively, a continuously operating low level of pressurization of the appropriate spaces can be incorporated as part of the normal ventilation arrangement for the building and an increased level of pressurization is then brought into operation in an emergency. This is called a *two-stage system.*

The two-stage system is generally regarded as preferable because with it some measure of protection is always operating and, therefore, any smoke spread in the early stages of a fire will be prevented. Additionally, any equipment, part of which is in continuous use, will be readily included in a satisfactory maintenance check. Whichever system is chosen for any particular building it must be understood that all the pressurized spaces (i.e. staircases, lobbies, corridors, as the case may be) must be pressurized in case of an emergency. It is not satisfactory to pressurize only those spaces in the immediate vicinity of the fire, nor is it acceptable to pressurize the spaces on the fire floor only. The system should be designed to develop the following

Table 5.2 *Design pressure differentials*

Building description	Pressure differentials (Pa)	
	Full system capacity for emergency operation	*Reduced system capacity for use in normal conditions when system is designed to run continuously*
Building height less than 25 m	50	8
Building height 25 m to 100 m	50	15
Building height above 100 m	50	25
Underground buildings	50	8

50 Pa = 50 N/m² = 0.2 in (5 mm) w.g. = 1/2000 atmosphere.

overall pressure differentials (Table 5.2) between the pressurized and fully vented unpressurized spaces, e.g. between staircases and office accommodation (in an office building).

For a single-stage system, Column 2 (for emergency operation figures) should be used; and for a two-stage system, Columns 2 and 3 should be used for emergency and normal operations respectively.

It is possible that in some small buildings an emergency pressurization level of 25 Pa may be adequate, and equally in a very exposed site the 50 Pa level may need to be reconsidered. However, it is suggested that a pressurization level of 60 Pa should not be exceeded, otherwise it may prove difficult to open a door leading into a pressurized space.

2. *Additional force required to open a door.* When a pressurization level of 50 Pa is used, the additional force which has to be exerted at the door handle in order to open it against this pressure is 40 N (assuming the door is 2 m high and 800 mm wide).

$$(40 \text{ N} = 9.0 \text{ lb wt} = 4.1 \text{ kg wt})$$

For a pressurization level of 60 Pa the additional force required to open this size door would be 48 N.

$$(48 \text{ N} = 10.8 \text{ lb wt} = 4.9 \text{ kg wt})$$

If the major design features of the pressurization system have been decided, the design sequence is now continued by considering the details of the building and the proposed system in the following order:

The air supply needed

The air supply which must be fed into any enclosed space in order to develop the required pressure differential is determined by rate at which air leaks out of that space. The relation between the pressure differential, the air flow and the cross-sectional area of the gap(s) through which leakage occurs is given by the equation:

$$Q = K \times A \times P^{1/N} \quad \text{(see Fig. 5.13)} \tag{5.1}$$

Where Q = the air supply into the space
A = cross sectional area of the leakage paths out of the space
P = pressure differential
K = a constant incorporating the discharge coefficient of the leakage paths, and the relationship between the units used
N = an index which can vary between 1 and 2.

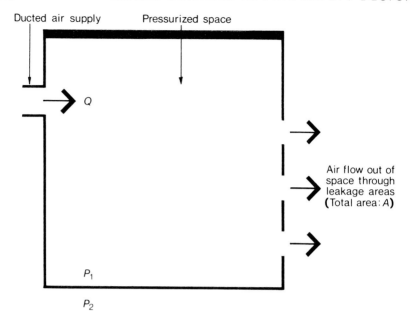

Fig. 5.13 *How an excess pressure is developed in a pressurized space.*
$P_1 = inside\ pressure$ *Pressure difference*
                                            $P = P_1 - P_2$
$P_2 = outside\ pressure$ *and thus air supply*
                                            $Q = K \times A \times P^{1/N}.$

It is important to note that the volume of the space being pressurized does not enter into the calculations. The size of the pressurized space will only affect the calculation if a significant component of the leakage out is through the building fabric, this leakage being proportional to the area of the enclosing surfaces. In the sort of construction which is normally used for protected escape routes, the leakage through the enclosing surface is not usually significant. However, it might in some circumstances be important, for instance if unrendered blockwork is used, and the design calculation must consider this possibility. In order to use the equation given above, values for K and N must be assigned. The value of K will vary according to the units used.

When Q is expressed in m^3/s
          A is expressed in m^2
          P is expressed in N/m^2 or Pa,
Then, $K = 0.827$,
Thus: $Q = 0.827 \times A \times P^{1/N}$ (5.1a)

When Q is expressed in l/s,
        A is expressed in m^2,
        P is expressed in N/m^2 or Pa,
Then, $K=827$,
Thus:
$$Q=827 \times A \times P^{1/N} \tag{5.1b}$$
When Q is expressed in ft^3/min
        A is expressed in in^2,
        P is expressed in in. w.g.
Then, $K=17.9$,
Thus:
$$Q=17.9 \times A \times P^{1/N} \tag{5.1c}$$

$$
\begin{bmatrix}
1\ \text{ft}^3/\text{min} = 0.472\ \text{l/s} & = 0.000472\ \text{m}^3/\text{s} \\
1\ \text{l/s} \quad\ \ = 0.001\ \text{m}^3/\text{s} & = 2.12\ \text{ft}^3/\text{min} \\
1\ \text{m}^3/\text{s} \quad = 2120\ \text{ft}^3/\text{min} & = 1000\ \text{l/s} \\
1\ \text{m}^3/\text{min} = 0.166\ \text{m}^3/\text{s} & = 35.3\ \text{ft}^3/\text{min} \\
1\ \text{m}^3/\text{h} \quad = 2.77 \times 10^{-4}\ \text{m}^3/\text{s} & = 0.59\ \text{ft}^3/\text{min}
\end{bmatrix}
$$

(See page 138 for pressure conversion.)

The value of N to be used will depend on the size of the individual leakage areas. For the large cracks around doors and for other large leakage areas the value of N should be 2. Thus, in the equation the quantity $P^{1/2}$ will be used.

For small openings, such as the cracks round windows, the value of 1.6 is more appropriate. Thus, in the equation the quantity $P^{1/1.6}$ or $P^{0.625}$ will be used. Values of $P^{1/2}$ and $P^{1/1.6}$ for values of P up to 60 Pa are given in Table 5.3.

Rules for adding leakage areas together

When considering the air leakage out of a pressurized space it is almost certain that there will be more than one place at which the air is escaping. It is part of the design procedure for all these leakage paths to be identified, and the problem arises as to how the several separate leakage areas are to be added together, so that a total effective value can be reached. The leakage paths can be in parallel with one another, or in series, or they can comprise a combination of series and parallel paths. The rules for calculating the total effective path are different for each of the above types and are explained in the following examples:

1. *Leakage paths in parallel.* This example occurs quite frequently, when the air leaks out of a pressurized space through several separate doors, each of these doors opening directly into an unpressurized space (Fig. 5.14). If A_1, A_2, A_3 and A_4 are the leakage areas of the cracks round the four doors indicated in Fig. 5.14, then the total effective leakage area A_T out of the pressurized space is given by:

$$A_T = A_1 + A_2 + A_3 + A_4 \tag{5.2}$$

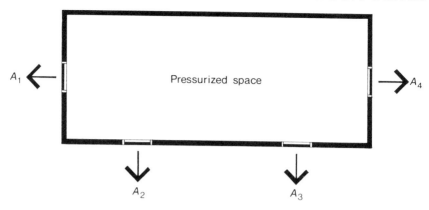

Fig. 5.14 *Leakage paths in parallel.*

2. *Leakage paths in series.* This condition arises when the air leaking out of a pressurized space passes through several individual leakage paths in succession; for instance, if air from a pressurized staircase leaks into a lobby, then out of the lobby into a corridor and then into another space before finally reaching the unpressurized space.

Table 5.3   *Values of $P^{1/N}$ for $N = 1.6$ and $N = 2$*

P	$P^{1/2}$	$P^{1/1.6}$	P	$P^{1/2}$	$P^{1/1.6}$	P	$P^{1/2}$	$P^{1/1.6}$
1	1.0	1.0	21	4.6	6.7	41	6.4	10.2
2	1.4	1.5	22	4.7	6.9	42	6.5	10.35
3	1.7	2.0	23	4.8	7.1	43	6.55	10.5
4	2.0	2.4	24	4.9	7.3	44	6.6	10.65
5	2.2	2.7	25	5.0	7.5	45	6.7	10.8
6	2.4	3.1	26	5.1	7.7	46	6.8	10.95
7	2.6	3.4	27	5.2	7.85	47	6.85	11.1
8	2.8	3.7	28	5.3	8.0	48	6.9	11.25
9	3.0	3.9	29	5.4	8.2	49	7.0	11.4
10	3.15	4.2	30	5.5	8.4	50	7.1	11.55
11	3.3	4.5	31	5.6	8.55	51	7.15	11.65
12	3.5	4.7	32	5.65	8.7	52	7.2	11.8
13	3.6	5.0	33	5.7	8.9	53	7.3	11.95
14	3.7	5.2	34	5.8	9.05	54	7.35	12.1
15	3.9	5.4	35	5.9	9.2	55	7.4	12.25
16	4.0	5.65	36	6.0	9.4	56	7.5	12.4
17	4.1	5.9	37	6.1	9.55	57	7.55	12.5
18	4.2	6.1	38	6.2	9.7	58	7.6	12.65
19	4.35	6.3	39	6.25	9.85	59	7.7	12.8
20	4.5	6.5	40	6.3	10.0	60	7.75	12.9

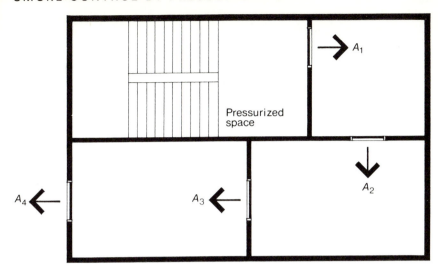

Fig. 5.15 *Leakage paths in series.*

If, in the case illustrated in Fig. 5.15, A_1, A_2, A_3 and A_4 are the leakage areas of the cracks round each of the doors shown, then the total effective leakage area from the staircase to the final unpressurized space is given by:

$$\frac{1}{A_T^2} = \frac{1}{A_1^2} + \frac{1}{A_2^2} + \frac{1}{A_3^2} + \frac{1}{A_4^2} \tag{5.3}$$

or:

$$A_T = \left(\frac{1}{A_1^2} + \frac{1}{A_2^2} + \frac{1}{A_3^2} + \frac{1}{A_4^3}\right)^{-1/2} \tag{5.4}$$

In the context of pressurization it is, in fact, unusual to have more than two leakage paths in series and in this case the total effective leakage area will be:

$$A_T = \frac{A_1 \times A_2}{(A_1^2 + A_2^2)^{1/2}} \tag{5.5}$$

Note: The calculations given in Equations 5.3 to 5.8 only strictly apply to leakage paths for which the value of N in Table 5.3 is 2 (i.e. for doors, cracks, or large openings). However, they may be used as an approximate calculation when windows (for which $N = 1.6$) form part of a series leakage path.

3. *Leakage paths in which series and parallel paths are both present.* This condition can occur when the pressurizing air (say, from a staircase) passes first through a small lobby, then into a larger space from which several doors

open. The example below shows such a situation and the method of calculation is explained.

In the example illustrated in Fig. 5.16 the only pressurized space is the staircase (none of the other spaces shown have an independent supply of air). The design requirement is to calculate the effective leakage area out of the staircase which results from air passing through the seven doors shown.

Start with the *series* path formed by space L_3. The effective sum of paths A_3 and A_4 will be called $A_{3/4}$, and,

$$A_{3/4} = \frac{A_3 \times A_4}{(A_3^2 + A_4^2)^{1/2}} \tag{5.6}$$

This effective sum ($A_{3/4}$) is now in *parallel* with the other three doors leading out of space L_2. Hence: The total leakage area out of space L_2 (which we will call $A_{7/3}$) is given by:

$$A_{7/3} = A_7 + A_6 + A_5 + A_{3/4} \tag{5.7}$$

This effective total leakage area out of space L_2 is in *series* with the two doors which lead through space L_1.

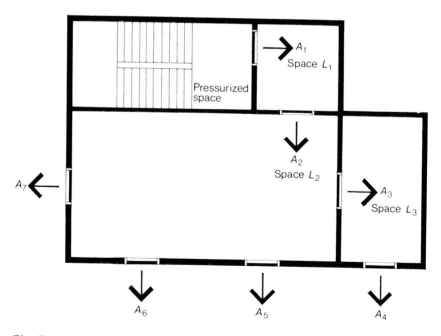

Fig. 5.16   *System with leakage paths in parallel and series.*

So, the effective leakage area from the pressurized staircase to the unpressurized spaces which lie outside spaces L_2 and L_3 (which we will call $A_{1/7}$) is given by:

$$A_{1/7} = \left(\frac{1}{A_1^2} + \frac{1}{A_2^2} + \frac{1}{A_{7/3}^2} \right)^{-1/2} \tag{5.8}$$

This may seem to be a complicated formula but the simple statement of the procedure is:

(*a*) Start at the low pressure end of the leakage paths;
(*b*) Combine each group of leakage paths by the appropriate formula to find the effective leakage area for that group;
(*c*) Work back successively towards the pressurized space.

Having stated the rules for adding the leakage areas it is now necessary to consider the actual values of the leakage areas of the various items found in a building.

Leakage areas for various components
In designing a pressurization system it is necessary to identify the several leakage paths through which air can escape from the pressurized space, and then to assess their size. This involves making assumptions about the area of the cracks round doors, windows and other places where air will escape, such as air bricks or other ventilation openings.

1. *Doors.* In general, the closed door is the most important place for air to leak away, and Table 5.4 gives typical values for the leakage area for the several types of door likely to be found as the closure to a pressurized space.

Table 5.4: *Typical leakage areas for four types of door*

Type of door	Size	Crack length (m)	Leakage area (m²)
Single leaf door in rebated frame opening into a pressurized space	2 m high 800 mm wide	5.6	0.01
Single leaf door in rebated frame opening outwards from a pressurized space	2 m high 800 mm wide	5.6	0.02
Double leaf door, with or without centre rebate.	2 m high 1.6 m wide	9.2	0.03
Lift landing door	2 m high 2 m wide	8	0.06

For doors smaller than the above sizes, the leakage areas given should be used. For larger doors, the leakage area should be increased in direct proportion to the increase in crack length. For instance, a single leaf door 2 m high and 1.2 m wide in a rebated frame, opening into a pressurized space, will have a leakage area of:

$$\frac{6.4}{5.6} \times 0.01 \text{ m}^2 = 0.0114 \text{ m}^2 \quad \text{(i.e. an increase of } 14\%)$$

The air leakage through a lift landing door will depend also on the air leakage out of the lift shaft, and the calculation of the effective leakage for a lift landing door is described in a later section.

Using the leakage areas given above and the expression for calculation given in Equation 5.1, the following values of air leakage past closed doors are obtained for the pressure differentials most commonly required for the design of a pressurization system (Table 5.5).

Table 5.5 *Air leakage data for doors (leakage expressed in m^3/s)*

Door type	Leakage area m^2	Pressure differential in Pa					Value of N
		8 Pa	15 Pa	20 Pa	25 Pa	50 Pa	
1. Single leaf opening into a pressurized space	0.01	0.0234	0.0320	0.0370	0.0413	0.0585	2
2. Single leaf opening outwards from a pressurized space	0.02	0.0468	0.0640	0.0740	0.0827	0.117	2
3. Double leaf	0.03	0.070	0.096	0.111	0.124	0.175	2
4. Lift landing door	0.06	0.14	0.192	0.222	0.248	0.351	2

2. *Windows.* In many buildings the pressurized space(s) will not be on an external wall and consequently will not have any window opening. However, there may be circumstances where an openable window opens out of a pressurized space and for this reason typical leakage data for windows are given below. Unlike doors, the sizes of windows will vary considerably and for this reason the leakage areas given in the Table below are for unit length of crack. In determining the leakage round an openable window, the total length of crack must be measured and this must be multiplied by the appropriate value for unit length given in Table 5.6.

TABLE 5.6   *Air leakage data for windows (expressed in m³/s per metre crack length for given pressure differentials in Pa)*

Window type	Crack area per metre length (m^2/m)	Pressure differential in Pa					Value of N
		8 Pa	15 Pa	20 Pa	25 Pa	50 Pa	
Pivoted	2.55 $\times 10^{-4}$	0.78 $\times 10^{-3}$	1.14 $\times 10^{-3}$	1.37 $\times 10^{-3}$	1.58 $\times 10^{-3}$	2.42 $\times 10^{-3}$	1.6
Pivoted and weather stopped	3.61 $\times 10^{-5}$	0.11 $\times 10^{-3}$	0.16 $\times 10^{-3}$	0.20 $\times 10^{-3}$	0.22 $\times 10^{-3}$	0.35 $\times 10^{-3}$	1.6
Sliding	1.00 $\times 10^{-4}$	0.31 $\times 10^{-3}$	0.45 $\times 10^{-3}$	0.54 $\times 10^{-3}$	0.62 $\times 10^{-3}$	0.95 $\times 10^{-3}$	1.6

Table 5.6 also gives the air leakage past openable windows per metre crack length for a selection of pressures which are most commonly met in design calculations. The air leakage for a given window is the total measured crack length multiplied by the values per metre obtained from the Table.

When it is necessary to find the leakage rate for other pressures attention is drawn to the value of N. The formula for making such a calculation is:

$$Q = 0.827 \times A \times P^{1/1.6} \qquad (5.9)$$

where Q is the air flow (m³/s), A is the total leakage area of the window crack (m²), and P is the pressure differential (Pa or N/m²).

3. *Lift landing doors.* When a door opening into a lift shaft is part of the air leakage path out of a pressurized space (Fig. 5.17), a special calculation is required. The effective leakage area cannot be regarded as simply the leakage area of the lift door (as given in Table 5.4) because the air leaks away from the lobby via the intermediate space of the lift shaft. In this case there is a combination of parallel and series leakage paths. Air from all the pressurized lobbies will flow into the lift shaft, so that all the lift doors are in parallel and the effective total of these parallel paths are in series with the leakage out of the lift shaft at the top vent (or of any other openings out of the lift shaft). In order to calculate the air flow past a lift door from a pressurized lobby, the following formula should be used:

$$Q = 0.827 \times \frac{A_1 \times A_2}{(A_1^2 + A_2^2)^{1/2}} \times P^{1/2} \qquad (5.10)$$

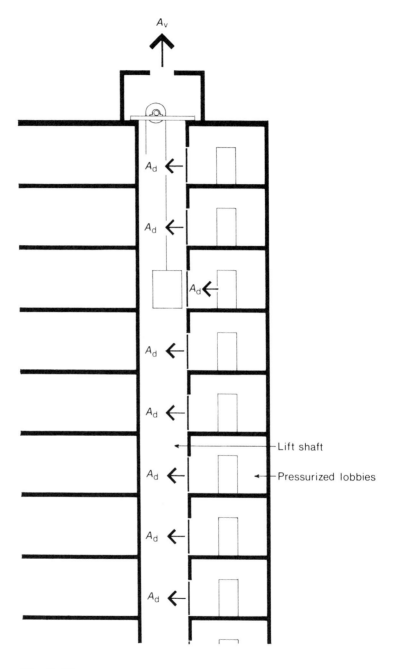

Fig. 5.17 *Diagram of lift landing door leakage – leakage areas* (A_d) *in parallel, and all in series with* A_v.

where Q = total air flow into lift shaft from all lobbies (m³/s)
             A_1 = total leakage area between all lobbies and the lift shaft (m²)
             A_2 = total leakage area between lift shaft and non-pressurized spaces
                 (m²)
             P = pressurization level of the lift lobby.

The required amount of air leakage from each lobby into the lift shaft can be determined by proportioning the total Q between all the lobbies in the ratio of each lobby's contribution to A_1. Usually each lobby will connect with each lift shaft by a similar door, and in this case the air leakage into each lift shaft from each lobby will be Q/n, and the value of A_1 will be $n \times A_L$, where n is the number of pressurized lobbies served by the lift, and A_L is the leakage area of one lift door – see Table 5.4.

A further simplification of the calculation is made possible, because it is quite usual to find that a vent of area 0.1 m² is provided at the top of a lift shaft. The new Code of Practice [1] specifies that the vent shall not be smaller than this. For a specified size of opening from the lift shaft to the outside air, the leakage from one lobby into the lift shaft can be calculated from the following equation.

$$Q_D = \frac{Q_L \times F}{n} \qquad\qquad (5.11)$$

where Q_D = air leakage from one lobby past one lift door
             Q_L = air leakage from an isolated lift door (value taken from Table
                 5.5 or calculated from $Q_L = 0.0496 \times P^{1/2}$, where P is the
                 pressurization level of the lobby)
             F = factor depending on vent size in lift shaft and taken from
                 appropriate column of Table 5.7
             n = number of pressurized lobbies opening into the lift shaft.

The significance of the three different vent sizes given in Table 5.7 is as follows:

(a) The figures in Column 2 (vent size 0.1 m²) should be used if the only opening from the lift shaft to an unpressurized space is of this size, and as will be understood, this is the normal design situation.

(b) The figures in Column 3 (vent size 0.16 m²) should be used if the total vent size adds up to this value or if, in addition to a vent size of 0.1 m² at the top of the lift shaft, there is one lift door leading to an unpressurized lobby. It will also be used in connection with calculations used when a door in one lobby is open, to be described later.

(c) The figures in Column 4 (vent size 0.22 m²) should be used if the total area of the vents out of the lift shaft add up to this value, or if in addition to a vent size of 0.1 m² at the top of the shaft, there are two doors in the same

TABLE 5.7   *Values of factor F for various vent sizes*

Number of floors served (n)	Value of F for vent size		
	0.1 m²	0.16 m²	0.22 m²
1	0.86	0.94	0.96
2	1.28	1.60	1.764
3	1.46	1.99	2.32
4	1.54	2.22	2.70
5	1.58	2.35	2.96
6	1.61	2.44	3.13
7	1.62	2.49	3.25
8	1.63	2.55	3.33
9	1.64	2.56	3.40
10	1.645	2.58	3.44
12	1.65	2.60	3.51
14	1.655	2.62	3.53
16	1.66	2.63	3.57
above 16	1.66	2.66	3.66

shaft opening into unpressurized lobbies. It will also be used if it is necessary to calculate air flows when doors in two lobbies are open.

If the leakage paths of the lift shaft (vent sizes) are different from any of the three values given in Table 5.7 above, then new values of F must be calculated using the following formulae:

$$F = \frac{A_t}{A_d} \tag{5.12}$$

where

$$A_t = \left[\left(\frac{1}{A_v} \right)^2 + \left(\frac{1}{n \times A_d} \right)^2 \right]^{-1/2} \tag{5.13}$$

$A_t =$ total effective leakage area of lift shaft
$A_d =$ leakage area of one lift door
$A_v =$ leakage area of vent (or openings into unpressurized spaces)
$n =$ number of floors served by the lift.

All of the calculations set out in this section deal with one lift, and it is assumed that the lift shaft is a protected shaft. A separate calculation must be made for each lift. When there are two or more lifts in a common shaft, it is sufficient for the purpose of calculation to treat each lift as being in its own single shaft, in which case the value of A_v used must be that which relates to each single lift (usually A_V for the large common shaft divided by the number of lifts in that shaft).

If a lift shaft connects together a series of pressurized lobbies, it should not also have a door which leads to an unpressurized lobby or space unless that lobby or space is not part of an escape route nor has any door communicating with an escape route.

The calculations detailed here for air leakage past a lift door are quite complicated and may seem to be confusing. However, an explanation of their use is given in a worked example later in this chapter.

Other series and parallel leakage paths

Similar combinations of series and parallel paths may occur in other respects (for instance, leakage into service shafts) and the methods used for the lift shaft may be suitably adapted and used in such cases, provided all the spaces involved are protected structures. Where an intermediate space is not a protected structure, then it must not be assumed that this space will remain pressurized, and the method of assessment of the air flow requirements must be based on the simple equation in (5.1a) to (5.1c).

Toilet areas

A fairly common leakage path which is associated with large lobbies is through toilet rooms. These form a fairly special case because these spaces are often virtually sealed and rely for ventilation on the extract system required by regulation. For such cases the leakage rate due to them is either:

(*a*) The extract rate in m^3/s when the extract fan is running (or is intended to run throughout an emergency),

or,

(*b*) when the extract fan does not run in an emergency, calculated from:

$$Q_t = Q \times K_2 \qquad\qquad (5.14)$$

where Q_t = the leakage into the toilet (or other) space;
          Q = door leakage rate in m^3/s at the design pressurization taken from Table 5.5, or calculated from Equation 5.1a;
and K_2 = a factor depending on A_B/A_D, taken from Table 5.8;
where A_B = minimum cross-sectional area of extract branch duct in m^2 (this may be a duct cross-section or the balancing device at the orifice, or damper);
          A_D = door leakage area including area of any air flow grilles or large gaps for air transfer in m^2.

Note: The value of A_D including transfer grilles and/or air flow clearance must also be used to calculate the value of Q above when the leakage area is greater than the normal total area of cracks.

Table 5.8 *Values of K_2*

A_B/A_D	K_2
4 or more	1
2	0.9
1	0.7
0.5	0.45
0.25 or less	0.25

Sizing the required air flow

All of the likely leakage paths out of a staircase or a lobby have been discussed in the foregoing paragraphs and the design process can proceed when all have been identified and their size estimated. From a knowledge of the total leakage from the space, a calculation can be made of the air supply needed to maintain a pressure differential of the required level. It follows that any space to be pressurized must be so constructed that any leakage of air through the building fabric will be minimal.

If the construction is of concrete then it will probably be satisfactorily leakproof; but blockwork will probably need to be rendered or plastered to ensure that it is leakproof. Additionally, attention must be paid to joints between walls, or between walls, floors and ceilings, to ensure that no incidental leakage occurs at these places. This last precaution is likely to be particularly important if a system-built structure is involved.

In calculating the air supply needed for a pressurization system two major assumptions have to be made. These are:

(*a*) That the leakage areas of the components (doors, lift doors and windows) which have been used in the calculations will apply to the components concerned when the building is completed.

(*b*) That no unidentified leakage areas out of the pressurized spaces are present.

In the face of these two necessary assumptions, it is suggested that an allowance of 25% should be added to the calculated values of the required air supply. It should be emphasized that this addition is suggested to make allowance for uncertainties in the values of the leakage areas which have been assumed. It is *not* intended as an allowance to take account of leakage in the supply ducting. The installer must make his own assessment of the likely leakage in his ductwork, and make provision for this.

The calculated value of the air supply must be delivered *in toto* to the pressurized spaces concerned, and the approving authorities will have the power to require evidence that the actual air flow agrees with the calculated value. However, before finalizing the air flow requirements, there are two more important features to be considered and for which very specific rules are laid down in the new Code of Practice. These are:

(*a*) The effect of an open door to a pressurized space and the air flow which passes through that open door; and

(*b*) The way(s) by which the pressurizing air will escape from the building.

The next section will deal with these important points.

Large openings in general

Design pressurization cannot be maintained if large openings exist between pressurized areas and neighbouring spaces, and in these circumstances other aspects of smoke control may need to be considered. In such cases smoke can be prevented from flowing through the opening if a sufficiently high air velocity out of the space is maintained. Thomas [38] has shown that when the opening is permanent, i.e. not a door opening intermittently, the air egress velocity would need to be 3 m/s to 4 m/s, depending on the temperature expected from the fire. This in turn will depend on the fire load and the ventilation; for low loads the lower velocity may be sufficient, but for high fire loads the upper value will be necessary. To obtain these velocities through large openings will require large volumes of air, and smoke control under these circumstances may well be either impractical or uneconomic except for very special circumstances.

The open door

No escape route can be effective without doors giving access to it, and it is inevitable that these be open from time to time. The design of a pressurization system must, therefore, have regard to the fact that a door to a pressurized space may need to be open for short periods and the smoke control function must be maintained in this event.

In the last section it has been stated that although when a large opening is made between a pressurized space and the surrounding space, a pressure difference cannot be maintained, the protection against smoke can be obtained by ensuring that a reasonable air velocity out through the large opening is established. For an intermittent opening such as a door a lower air velocity than that suggested for the permanent opening can be used, and the value will depend on the position of the door. There are three major situations, as follows:

(*a*) When the staircase only is pressurized, with no intervening lobby, then the minimum air egress velocity through an open door is required to be 0.75 m/s. If the building has more than 20 storeys, then this egress velocity must be maintained when two doors on different floors are open.

(*b*) When the staircase and a lobby on each floor are independently pressurized, then a minimum air egress velocity of 0.5 m/s is required through an open lobby door, provided there is another closed door leading to the staircase. If the building has more than 20 storeys, this air velocity must be maintained when lobby doors on two floors are open. However, this condition is waived provided condition (*c*) is satisfied and provided a pressure differential of 50 Pa is maintained across the closed door leading to the staircase.

(*c*) When the staircase and a lobby on each floor are independently pressurized, then a minimum air egress velocity of 0.7 m/s is required when two lobby doors on one floor are open. (This means a staircase-to-lobby door and a lobby-to-accommodation door are open.) This egress velocity can be at either of the open doors, and when the building is of more than 20 storeys the required velocity must be obtained when two lobby doors on two floors are open.

In general terms, the air flow requirement through an open door out of a pressurized space is a velocity of 0.5 m/s when there is another closed door leading to the staircase or 0.7 m/s when there is no closed door protecting the staircase. These values of air flow can only be achieved when there is a satisfactory degree of air leakage from the accommodation part of the building, and the three generalized examples given above assume that the lobby or staircase door opens directly into the accommodation of the building on each floor, and that this accommodation is open-planned and not sub-divided.

When this is not the case, and the stairs and/or lobbies lead into corridors (or other spaces), and then to a sub-divided floor, the stipulation needs to be made that the specified air flows out of the staircase and lobby doors must be measured when other doors in the partitioned space are open; otherwise there will not be a sufficient amount of leakage out of the building to allow these air flows to develop. As part of the design calculations, the designer should show the value of the egress velocity which will apply when doors as specified above are open. An exact calculation of this velocity is difficult and the very considerable work involved would not be justified in this context because of the uncertainties caused by the several assumptions which have to be made in the design calculations. An approximate calculation is, therefore, sufficient and a method of making this assessment is given in the Appendix to this Chapter.

If the required egress velocities are not achieved as a result of the initial design, then the air input values to the staircase must be increased until these open door requirements are satisfied. When this is necessary, additional permanent opening(s) must be placed in the staircase to prevent the pressure rising above 50 Pa when all the doors are closed. This additional leakage area can be closed by a counterbalanced flap valve so designed that it will only open when the pressure exceeds 60 Pa. The flap valve would usually be placed between a pressurized space and an internal space. An example of a flap valve is shown in Fig. 5.18.

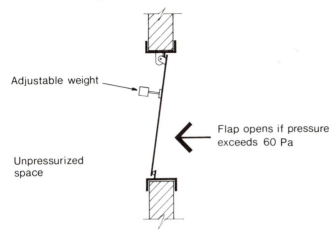

Adjustable weight

Flap opens if pressure exceeds 60 Pa

Unpressurized space

Fig. 5.18 *Pressure-limiting device.*

The area of the opening which the flap valve is arranged to close is calculated using the normal equation relating to pressure, air flow and leakage area which is given earlier in the Chapter (Equation 5.1). If Q_A is the *additional* air supply needed to satisfy the open doors requirement, A is the area of the flap valve and the design pressurization level is 50 Pa,

then,	$A = Q_A/5.85$ when A is in m² and Q_A is in m³/s.	(5.15)
or,	$A = Q_A/5850$ when A is in m² and Q_A is in l/s.	(5.16)
or,	$A = Q_A/8$ when A is in in² and Q_A is in ft³/min	(5.17)

5.5.2 General requirements regarding supply of pressurizing air

There are necessarily some special points about the distribution and supply of the pressurizing air which should be observed.

1. Staircases. When several staircases are pressurized in each building a separate pressurization system should be provided for each staircase. The air supply to a pressurized staircase must be evenly distributed throughout the whole height of the staircase and should be ducted up or down the staircase with supply grilles at intervals not exceeding three storey heights between adjacent grilles. The outlets must be arranged and balanced so that equal quantities of air flow from each outlet grille. This means that a single supply entry point is not acceptable unless the building has 3 storeys or less.

2. Lobbies. In general, lobbies may be pressurized using a common fan and duct system, provided suitable balancing arrangements ensure that the correct air supply is provided to each lobby. However, the duct and terminal designs must be so arranged that when the pressure in one or two lobbies is reduced because of open doors, the reaction on the air supply to other lobbies will be minimal. If the scheme consists of a staircase and associated lobbies then a common pressurization fan may be used but with two duct systems, one for the staircase and one for the lobbies. If more than one staircase has access to a common lobby, then separate pressurization systems must be used for each staircase, but all or any of the staircase pressurization systems may be used to supply the lobbies, provided a duct run is used for the lobbies which is separate from that used for the staircase(s).

3. Door clearance(s). At the design stage for a pressurization scheme it is necessary to make assumptions about the air leakage past doors, windows and other building components, so that the size required for the fan and ductwork can be specified.

It is absolutely essential, therefore, to ensure:

(*a*) That notwithstanding the information given earlier (in Tables 5.4 and 5.6), the leakage areas assumed in the actual design calculations are reasonable for the particular items (doors, windows, etc.) to be used in the building.

(*b*) That when fitted, these items do conform to the leakage assumptions made.

This last point is most important, and a common difficulty arises in connection with the clearance at the bottom of a door, If, for example, because of a change in the thickness of the floor covering proposed, a large gap is left at the bottom of a door, this would not be regarded as important from the fire resistance point of view, but it could have a major effect on the operation of a pressurization system. Such a change in floor covering could affect all the doors in a building.

4. Fire dampers in the ductwork. Since a pressurization system must continue to operate for the duration of a fire, the ductwork associated with the system should be so positioned in the building that the need for fire dampers is avoided. Ducts which are contained in protected shafts do not normally need to be fitted with fire dampers.

5.5.3 The escape of the pressurizing air from the building

General

The last feature in the design consideration of a pressurization scheme is an examination of how and where the pressurizing air will leak out of the building, since this is an important part of the concept and it may be necessary to make special arrangements for this air release. The principle of pressurization requires that either a stated pressure differential is developed across a closed door which gives access to the pressurized space from the unpressurized accommodation or a specified air flow will be established through the door opening if the door is open. It further postulates that an air flow pattern shall be established across the unpressurized space so that the general air movement will always take the smoke and hot gases away from the door leading to the escape route. This condition will not be fulfilled unless the air leaking or flowing past the door is able to escape from the accommodation space to the open air by means of a low resistance path which is remote from the door(s) leading to the pressurized space.

Methods of arranging for the escape of pressurized air

There are four possible ways in which the escape of the pressurizing air can be achieved. These are:

(*a*) By window leakage.
(*b*) By specially provided vents at the building periphery.
(*c*) By the provision of vertical shafts.
(*d*) By mechanically operated extraction.

These are discussed below, and in this discussion Q_N is equal to the net volume of pressurizing air flowing into the floor, and the value (in m^3/s) for the open door condition should be taken for this purpose.

1. Window leakage. When a building has openable windows on every floor it is possible that the leakage through these will be sufficient to allow satisfactory venting of the pressurizing air. Table 5.9 shows the total length of window cracks which is satisfactory for this purpose. In assessing the available total length of cracks, one face of the building should be discounted because of possible adverse wind conditions. If the window leakage is not evenly distributed around the external wall, then the side with the largest area of window leakage should be discounted.

2. Specially provided vents at the building periphery. When the building is sealed or insufficient openable windows are available the necessary leakage can be provided by special vents in the external walls which should preferably be evenly distributed round the building. The minimum value of the total effective vent opening area per floor should be $Q_N/2.5$ m^2. For the case of one open door through which an air velocity of 0.5 m/s is required the minimum vent opening area would be 0.32 m^2 (3.5 ft^2). This is a minimum area and it

Table 5.9 *Minimum total length of window cracks (per floor) for satisfactory venting of pressurizing air*

Window type	Recommended crack length in metres	
	For any value of Q_N	*For 0.5m/sec through one door* $Q_N = 0.8 \ m^3/sec$
Pivoted, no weather stripping	$1200 \times Q_N$	960
Pivoted and weather stripped	$3000 \times Q_N$	2400
Sliding	$8300 \times Q_N$	6640

is advantageous to use larger areas if this is possible. Again the possible effect of an adverse wind must be considered, and in assessing the effective area of venting, one side of the building should be discounted, and if the vent distribution is uneven then the side with the largest vent area should be discounted.

Vent opening design: It is unlikely that vents in the form of permanent openings will be acceptable during the normal use of the building and so a vent design must be used which ensures that the vents are open when the emergency pressurization system starts operating. The necessary features of the vent opening(s) (Fig. 5.19) are:

(*a*) the vent closure should be normally held (or should rest) in the closed position;

(*b*) when the emergency pressurization system operates the vent closure should be released so that the pressurizing air is free to escape without having to develop any appreciable pressure to do so;

(*c*) the vent closure should be capable of being closed by the action of adverse wind on any particular face of the building;

(*d*) if automatically controlled venting is proposed, then it is preferable that the venting should take place on the fire floor only. On all other floors, although the pressurization on those floors is active, the vents should remain closed. However, design calculations must assume venting takes place on all floors.

3. Venting by vertical shafts. The use of vertical shafts can be a satisfactory way of releasing the pressurizing air from the accommodation of an otherwise sealed building. The minimum size of shaft and vent which should be used are as follows:

Net vent area per floor (accommodation into shaft)=

$$A_v = \frac{Q_N}{2}\ m^2 \tag{5.18}$$

when the air velocity through one open door is 0.5 m/s then,

$$Q_N = 0.8\ m^3/s \quad \text{and} \quad A_v = 0.4\ m^2 \tag{5.19}$$

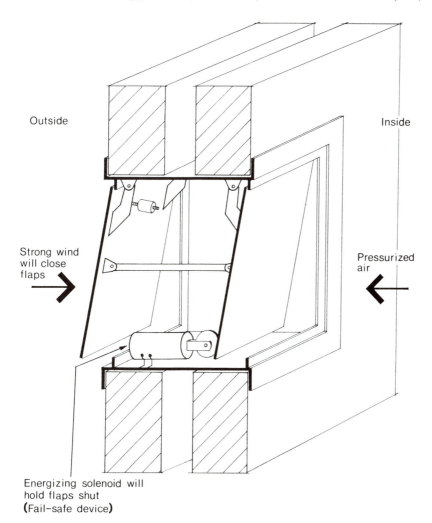

Outside

Inside

Strong wind
will close
flaps

Pressurized
air

Energizing solenoid will
hold flaps shut
(Fail–safe device)

Fig. 5.19 *Possible design for vent to release pressurized air at external wall.*

The shaft size and the top vent size should not be smaller than the vent area on one floor. These sizes are total values and can be obtained by the use of smaller shafts distributed in the accommodation space.

If the vents into the vertical shafts from the accommodation floor are to be *permanently open* then either a separate shaft must be provided for each floor or a shunt duct system to avoid smoke and fire-spread between floors must be used. Fire dampers cannot be used in a shaft provided for the venting of the pressurizing air.

However, the use of permanently open vents is unlikely to be regarded as either practical or desirable, and an arrangement which uses an *automatically opening* vent is to be preferred. In this case, a common shaft is used for all floors and the vent on each floor is normally closed by a fire-resisting closure, which, when the emergency pressurization system is brought into operation, opens on the fire floor only. On all other floors the vent remains closed.

The use of vertical shafts for venting the pressurizing air has at least two quite significant advantages:

(*a*) The difficulties due to adverse wind conditions can be readily avoided because a vertical discharge (with a suitable cowl) can be arranged.

(*b*) The stack effect of a vertical shaft full of hot smoke can give distinct assistance to the pressurization system [35].

4. The use of mechanical extraction: The release of the pressurizing air by using mechanical extraction is a satisfactory method provided suitable precautions are incorporated.

The important requirements are:

(*a*) the extract rate per floor should be not less than Q_N m^3/s.

(*b*) the system (ducts and fans) must be capable of withstanding high temperatures (500°C) for a reasonable period of time.

(*c*) smoke and fire must not be able to spread from floor to floor.

These requirements can be satisfied by having a separate extract system for each floor.

Alternatively, if a system common to all floors is used, then in the case of fire the ducts on all floors not affected by fire must be closed off by a fire-resisting damper. The damper on the fire floor remains open. For either system the ductwork must be constructed to the appropriate standard of fire resistance, and when the common system is used, the extract rate must be such that an extraction of Q_N can be maintained on the fire floor.

It is worth repeating that the design of the system which allows the pressurizing air to escape from the building is most important. The Table overleaf (Table 5.10) summarizes the methods and makes suggestions for their use.

Table 5.10 *Suggestions for choice of venting system*

Building layout	Windows	Ventilation	Venting system	
			Main methods	Additional or alternative methods (if required)
Open plan	Openable, not weather stripped	Natural	Natural leakage (window leakage)	—
	Openable, weather stripped	Natural or mechanical	Natural leakage or vertical shafts	Peripheral vents or mechanical extract
	Sealed	Mechanical	Peripheral vents or vertical shafts	Mechanical extract
Floors partitioned	Openable, not weather stripped	Natural	Natural leakage (window leakage)	—
	Openable, weather stripped	Natural or mechanical	Natural leakage or peripheral vents	Vertical shafts or mechanical extract
	Sealed	Mechanical	Peripheral vents	Vertical shafts or mechanical extract

5.5.4 Design procedure to be followed when designing a pressurization system for a building

(*a*) Consider the proposals for the building and indicate changes in layout which will be possible or necessary if pressurization is to be used.

(*b*) Identify the spaces to be pressurized and consider any possible interaction between pressurized and unpressurized space.

(*c*) Decide whether the system is to be single or two-stage and select the levels of pressurization to be used for emergency operation and, if appropriate, for reduced capacity operation (Table 5.2).

(*d*) Identify all the leakage paths through which air can escape from the pressurized space(s) and determine the rate of air leakage through each for the appropriate pressure differential.

(*e*) Total all the air flows out of each pressurized space and increase this total by 25% in accordance with the reason given in the text. This will give the air supply needed for each pressurized space.

(*f*) The air velocity through an open door should be estimated using the appropriate procedure set out in Appendix A. If the requirements are not satisfied, the air supply proposed must be increased.

(*g*) The air supply in (*e*) and (*f*) above must be provided at the duct terminal (or terminals) in each pressurized space. The positions of the duct terminals should be discussed with the architect and the appropriate authorities.

(*h*) The fan capacity and duct sizes must be decided by a competent engineer after due consideration of any additional requirements. The position of the intake grilles must be agreed with the architect and any special protection to the installation specified.

(*i*) The escape of the pressurizing air from the building should be considered and the appropriate method of venting specified.

(*j*) The operation of the system must be considered and the position of smoke detectors (if required) specified.

(*k*) A note of the leakage areas assumed must be given to the architect, reminding him that these areas must be achieved in the finished building.

(*l*) A measurement procedure should be specified so that the satisfactory operation of the installation in the completed building can be established.

A worked example is given in Appendix B to this Chapter.

References

1. British Standard B.S. 5588 (1978). *Code of Practice for Fire Precautions in the Design of Buildings. Part 4. Smoke Control in Protected Escape Routes using Pressurization.* British Standard Institution, London.
2. Malhotra, H. L. (1967). Movement of smoke on escape routes. Part 3. Effect of permanent openings in external walls. Fire Research Note No. 653, Fire Research Station, Borehamwood, England.
3. Wilkinson, J. (1969). The Worthing A.A.C. system, automatic air flow control system for escape routes in multi-storey blocks of flats. Paper 11, Fire Research Station Symposium No. 4, *Movement of smoke in escape routes in buildings.* Watford 1969. HMSO 1971.
4. Fire Protection Code for Buildings over 150 ft in Height, Interim. (1957). Amendment to Height of Building Act 1957. State Planning Authority of New South Wales, Australia.

5. Malhotra, H. L. and Millbank, N. (1964). Movement of smoke in escape routes and effect of pressurization. Results of some tests performed in a new department store. Fire Research Note No. 566, Fire Research Station, Borehamwood, England.

6. Butcher, E. G., Fardell, P. J. and Clarke, J. J. (1969). Pressurization as a means of controlling the movement of smoke and toxic gases on escape routes. Paper 5, Fire Research Station Symposium No. 4, *Movement of smoke on escape routes in buildings*. Watford 1969. HMSO 1971.

7. *Fire Research* 1969. Pressurization and the open door. Annual Report of Fire Research Station, 1969, p. 25. HMSO 1970.

8. Butcher, E. G., Fardell, P. J. and Clarke, J. J. (1969). Prediction of the behaviour of smoke in a building using a computer. Paper 9, Fire Research Station Symposium No. 4, *Movement of smoke on escape routes in buildings*. Watford 1969. HMSO 1971.

9. Tamura, G. T. (1969). Computer analysis of smoke movement in tall buildings. *American Society of Heating, Refrigeration and Air-Conditioning Engineers Trans.*, **75**, Pt. II, 81.

10. Barrett, R. E. and Locklin, D. W. (1969). A computer technique for predicting smoke movement in tall buildings. Paper 10, Fire Research Symposium No. 4, *Movement of smoke on escape routes*. Watford 1969. HMSO 1971.

11. Wakamatsu, T. (1975). Unsteady state calculation of smoke movement in an actually fired building. Paper 8, C.I.B. Symposium, *Control of smoke movement in building fires*, Watford 1975. Organized by Fire Research Station, Borehamwood, England.

12. Shannon, J. M. A. (1975). Computer analysis of the movement and control of smoke in buildings with mechanical and natural ventilation. Paper 9, C.I.B. Symposium, *Control of smoke movement in building fires*, Watford 1975. Organized by Fire Research Station, Borehamwood, England.

13. Appleton, I. C. (1975). A model of smoke movement in buildings. Paper 10, C.I.B. Symposium, *Control of smoke movement in building fires*. Watford 1975. Organized by Fire Research Station, Borehamwood, England.

14. Hobson, P. J. and Stewart, L. J. (1972). Pressurization of escape routes in buildings. Fire Research Note No. 958, Fire Research Station, Borehamwood, England.

15. National Fire Protection Association (1959). *Operation school burning*. Official report on tests conducted by the Los Angeles Fire Department, U.S.A.

16. Commonwealth Experimental Building Station (1972). Flame spread and smoke tests. C.E.B.S., R.F. No. 35, New South Wales, Australia.

17. Galbreath, M. (1969). Time for evacuation by stairs in high buildings. *Fire Fighting in Canada*, **13**, No. 1.

18. Tamura, G. T. and Wilson, A. G. (1968). Pressure differences caused by wind on two tall buildings. *American Society of Heating, Refrigeration and Air-conditioning Engineers Trans.*, **74**, Pt. II, 170.

19. Tamura, G. T. and Wilson, A. G. (1966). Pressure differences for a 9-storey building as a result of chimney effect and ventilation system in operation. *American Society of Heating, Refrigeration and Air-conditioning Engineers Trans.*, **72**, Pt. I, 180.

20. Tamura, G. T. and Wilson, A. G. (1967). Building pressures caused by chimney action in 3 high buildings. *American Society of Heating, Refrigeration and Air-conditioning Engineers Trans., 73*, Pt. II, 1.1–1.10.

21. Tamura, G. T. and Wilson, A. G. (1967). Building pressures caused by chimney action and mechanical ventilation. *American Society of Heating, Refrigeration and Air-conditioning Engineers Trans., 73*, Pt. II, 2.1–2.9.

22. Tamura, G. T. (1969). Computer analysis of smoke movement in tall buildings. *American Society of Heating, Refrigeration and Air-conditioning Engineers. Trans., 75*, Pt. II, 81.

23. Tamura, G. T. (1970). Analysis of smoke shafts for control of smoke movement in buildings. *American Society of Heating, Refrigeration and Air-conditioning Engineers Trans., 76*, Pt. II, 290.

24. Tamura, G. T. and Wilson, A. G. (1970). Natural venting to control smoke movement in buildings, via vertical shafts. *American Society of Heating, Refrigeration and Air-conditioning Engineers Trans., 76*, Pt. II, 279.

25. National Research Council of Canada (1970). Explanatory paper on control of smoke movement in high buildings. N.R.C. Report No. 11413, Ottawa, Canada.

26. Uniform Building Code, U.S.A. (1970). Smoke proof enclosures. Vol. 3, 1970. Housing section 3309.

27. Los Angeles Fire Department (1966). *Vertical Enclosures*. Report published by L. A. Fire Department 1966. U.S.A.

28. Los Angeles Fire Department (1970). *Mechanically ventilated smoke proof enclosures*. City of Los Angeles B.F.P. and P.A. Requirement No. 56, U.S.A.

29. Degenkolb, J. G. (1971). Smoke proof enclosures. *American Society of Heating, Refrigeration and Air-conditioning Engineers Journ., April, 33–8*.

30. Centre Scientifique et Technique du Batiment (1971). C.S.T.B. Report No. 70–4340. 28 May 1971.

31. Riou, J. (1972). Smoke extraction in buildings. A solution. *Fire International, 37*, 41–54.

32. Cabret, A. and Ferrie, M. (1969). Smoke protection of escape routes in buildings. Paper 6, Fire Research Station Symposium No. 4, *Movement of smoke in escape routes*. Watford 1969. HMSO 1971.

33. DeCicco, P. R., Cresci, R. J. and Correale, W. H. (1972). *Fire tests, analysis and evaluation of stair pressurization and exhaust in high rise office buildings*. Report published by Center for Urban Environmental Studies, Polytechnic Institute of Brooklyn, New York, U.S.A.

34. Koplon, N. E. (1973). *Report of the Henry Grady fire tests*. Report published by City of Atlanta Building Department, Atlanta, Georgia, U.S.A.

35. Butcher, E. G., Parnell, A. C. and Eastham, G. (1976). Smoke control by pressurization. *Fire Engineers' Journal, 36*, 103, 16–19.

36. Fire Check Consultants (1976). Fire set in a new building to test smoke control by pressurization. *Fire*, July 1976, 71–3.

37. Butcher, E. G., Cottle, T. H. and Bailey, T. A. (1971). Smoke tests in the pressurized stairs and lobbies of a 26-storey office building. *The Building*

Services Engineer, **39**, 206–10. Building Research Establishment, Current Paper CP 4/74.
38. Thomas, P. H. (1968). Movement of smoke in horizontal corridors against an air flow. Fire Research Note No. 723, Fire Research Station, Borehamwood, England. *Inst. Fire Engineers' Quarterly* 1970, **30**, 77, 45–53.

APPENDIX A   METHOD OF ESTIMATING AIRFLOW THROUGH AN OPEN DOOR

The procedure and methods given in this Appendix for estimating the air flow through an open doorway are not exact, but are sufficiently accurate for the purpose in hand, and in many cases represent a very substantial simplification in the calculation procedure. The uncertainties in the leakage resistance of actual systems do not justify a more sophisticated calculation procedure.

The calculations outlined below are based on an open floor plan and an infinite leakage from the periphery of the building. The velocity actually attained in an open doorway when practical venting requirements apply will be lower than that predicted and, therefore, the air velocity calculated by the method outlined should be multiplied by a factor of 0.6 in order to yield a better approximation to the actual velocity obtained in practice.

In all the calculations the open door is assumed to be single leaf and 1.6 m^2 in area.

5.A.1   When staircase only is pressurized

(*a*) For buildings of ten storeys or less, it is sufficient to assume that all the air input to the staircase will flow out of the open door.

(*b*) For buildings of more than ten storeys, the leakage area of all the other (closed) doors in the staircase (plus any other leakage areas) must be totalled and the proportion of the input air which will flow out of the 1.6 m^2 door area calculated.

Example: If there are 20 storeys and 21 double leaf doors leading out of the staircase, the total leakage area of the closed doors (with one open) will be $(21-1) \times 0.03 \text{ m}^2$. Hence, the proportion of input air which will flow out of the open door will be:

$$\frac{1.6}{(1.6 + (21-1) \times 0.03)} = 0.727$$

[*Note:* When the calculation of (*a*) or (*b*) above has been completed, the result must be multiplied by the factor 0.6.]

5.A.2 Staircase with lobby on each floor independently pressurized. One door open (lobby/accommodation door)

1. Lobbies not connected by lift shafts. The total air flow out of an open lobby/accommodation door will be the sum of:

(*a*) the air supplied to the lobby by the supply duct
(*b*) the air flow past a closed door from the staircase into the lobby.

To calculate (*b*) it is sufficient to assume that the design pressure in the staircase is maintained and that the pressure in the lobby with the open door falls to zero. (Hence the values in Table 5.5 can be used.)

2. Lobbies connected by one or more lift shafts. The total air flow out of an open lobby/accommodation door will be the sum of:

(*a*) the air supplied to the lobby by the supply duct
(*b*) the air flow past a closed door from the staircase into the lobby. Calculate this as indicated for 1(*b*) above
(*c*) the air flow out of each lift shaft past the closed lift door. Calculate this by assuming that the air flow *into* each lift shaft is still the value used in the design calculations. Then, approximately $\frac{1}{3}$ of this will flow past the closed lift door into the lobby with the lobby/accommodation door open. This will apply for *each* lift shaft.

If there are two lobbies in the building with an open door, then approximately $\frac{1}{4}$ of the total air flowing into each lift shaft will flow into each lobby.

[*Note:* When the calculations of *1* or *2* above have been completed, the result must be multiplied by the factor 0.6.]

5.A.3 Staircase with lobby on each floor independently pressurized. Two doors in same lobby open (one staircase/lobby and one lobby/accommodation door)

1. Lobbies not connected by lift shaft(s). The total air flow past the door between the lobby and staircase will be the sum of:

(*a*) the air supplied to the staircase by the supply duct
(*b*) all the air which will flow into the staircase past the closed door of all the other lobbies.

The value of (*b*) is calculated using the formula:

$$Q_{\mathrm{T}} = Q_{\mathrm{L}} \times \frac{A_{\mathrm{D}}}{A_{\mathrm{T}}} \times (n-1) \qquad (5.\mathrm{A}1)$$

where Q_{T} = air flow into staircase from all the lobbies which has closed doors.

Q_L = air supplied by duct to one pressurized lobby.

A_D = leakage area of the closed door between the lobby and staircase (assumed to be the same on all floors).

A_T = total leakage area through which air leaks out of each lobby.

$= A_D + A_L$, where

A_L = leakage area of each lobby used in the design calculations.

n = number of storeys having pressurized lobbies.

If two lobbies in the building have both doors open, then the air flowing through the staircase door to each lobby will be one half of the total of (a) and (b) above, and the factor $(n-2)$ must be substituted for $(n-1)$ in the equation above.

The air flow past the second lobby door (I.e. lobby/accommodation door) will be the sum of:

(a) the air supplied to the lobby by the ducted supply

(b) the air flowing out of the staircase through the open staircase lobby door (calculated as indicated above for one lobby or two lobbies with two doors open as appropriate).

2.   *Lobbies connected by lift shaft(s)*. The total air flow past the door between the lobby and staircases will be the sum of:

(a) the air supplied to the staircase by the supply duct

(b) all the air which will flow into the staircase past the closed door of all the other lobbies.

The value of (b) is calculated using the formula:

$$Q_T = Q_L \times \frac{A_D}{\left(A_D + A'_L + \dfrac{m}{n} \times A_E \times F\right)} \times (n-1) \qquad (5.A2)$$

where Q_T = air flow into staircase for all lobbies which have closed doors.

Q_L = air supplied by duct to one pressurized lobby.

A_D = leakage area of the closed door between lobby and staircase (assumed to be the same on all floors).

A_E = leakage area of lift entrance door (usually assumed to be 0.06 m²).

A'_L = leakage area of each lobby used in the design calculations *excluding the leakage area of the lift doors.*

n = number of storeys.

m = number of lift shafts opening into each lobby.

F = Factor listed in Table 5.7 for the appropriate number of storeys using, Column 3 or Column 4 of that Table according to whether one lobby or two lobbies in the building have two doors open. In the latter case the factor $(n-2)$ should be substituted for $(n-1)$ in the equation above.

Then the total air flow through the open staircase/lobby door will be equal to:

$$Q_S + Q_T$$

where Q_S = air supplied by duct to the staircase.

The air flow past the open door between the lift lobby and the accommodation, will be:

(a) the total air flowing into the lobby from the staircase

$$= Q_S + Q_T$$

(b) the air supplied by duct to each lobby

$$= Q_L$$

(c) the air flowing out of all the lift shafts past the closed lift door

$$= Q_A$$

The value of (c) is calculated using the formula:

$$Q_A = \frac{m}{3}\left(\frac{F \times Q_L \times A_E}{A_D + A'_L + (A_E \times F \times m/n)}\right) \qquad (5.A3)$$

[When two lobbies each have two doors open, the factor at the front of the above equation should be $m/4$, and Column 4 of Table 5.7 should be used for F.]

The total of (a), (b) and (c) above, i.e. the air flow past the open door between the lift lobby and the accommodation, can be written as:

$$\text{Total air flow past door} = Q_S + Q_L\left[\frac{A'_L + nA_D + ((m/n) + (m/3))A_E F}{A_D + A'_L + (m/n)A_E F}\right] \qquad (5.A4)$$

[*Note:* The volume air flows calculated for 1 and 2 above should be multiplied by the factor 0.6 for the reasons given earlier.]

3. *Air velocity past open door.* The calculations detailed in this Appendix give the volume rate of air flow through an open door. In general, this volume rate will be expressed in m³/s, but the requirement is for a specified velocity.

It can be assumed that the open door will be single leaf (or one leaf of a double leaf door) and that its area will be 1.6 m². Therefore, to express the air flow past the door as a velocity (in m/s) the values obtained for the volume flow must be divided by 1.6.

APPENDIX B   EXAMPLE OF DESIGN OF A PRESSURIZATION SCHEME

The example considered is a simple, relatively small, building which has a single staircase opening out into a small lift lobby on each floor (Fig. 5.20). Details of the building are:

Six storeys including ground floor.
Staircase has six single leaf doors to each lift lobby, and a double leaf door to open air at ground level.
Lift lobby has one lift door at each level and a double leaf door leading to accommodation at each level.
Lift shaft has a 0.1 m² vent at top, otherwise no leakage to open air.

In order to illustrate fully the design procedure, the twelve steps listed on pp. 161–162 (Chapter 5) will be considered.

(*a*) *Consider the proposals for the building and indicate changes in layout which will be possible or necessary if pressurization is to be used.*
No changes are necessary, but a more flexible layout could be considered because the staircase and lobby do not need to be on an external wall. Ducts to supply pressurizing air to staircase and lobby must be accommodated and space must be made for the pressurizing fan.

(*b*) *Identify the spaces to be pressurized and consider any interaction between pressurized and unpressurized spaces.*
The staircase and lift lobby will be independently pressurized. There is no likelihood of untoward interaction between pressurized and unpressurized spaces.

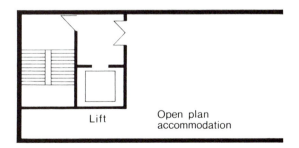
Lift   Open plan accommodation

Fig. 5.20 *Plan of building used for example of pressurization design.*

(c) *Decide whether the system is to be single or two-stage and select the levels of pressurization to be used for emergency operation and, if appropriate, for reduced capacity operation.*

A two-stage system will be chosen for this building with an emergency pressurization level of 50 Pa, and a reduced level for normal running of 8 Pa (since building height is less than 25 m).

(d) *Identify all the leakage paths through which air can escape from the pressurized space(s) and determine the rate of air leakage through each for the appropriate pressure differential.*

Staircase
    (i) Six doors to lift lobby. Since both lift lobby and staircase are pressurized to the same level there is no air movement across these doors.
    (ii) The only air leakage path out of the staircase is the double leaf door to open air at ground level.

Thus, air leakage out of staircase (from Table 5.4)$\simeq$0.03 m^2, air supply for normal pressurization level of 8 Pa$\simeq$0.070 m^3/s (from Table 5.5), and air supply for emergency pressurization level of 50 Pa$\simeq$0.175 M^3/s (from Table 5.5).

Each lift lobby has one double leaf door to accommodation and one lift door into lift shaft (staircase door has no air flow). Thus, double leaf door leakage area $\simeq$ 0.06 m^2 (Table 5.4).

Air leakage values from Table 5.5:

	8 Pa	50 Pa
Double leaf door	0.070 m^3/s	0.175 m^3/s
Lift landing door	0.14 m^3/s	0.351 m^3/s

(These values for lift landing door are uncorrected.)

If the lift shaft has a vent of 0.1 m^2 at the top and there are six floors, the value of the factor F needed to obtain the actual leakage from each lobby into the lift shaft = 1.61. (From Table 5.7, Column 2.) Thus, actual air flow into lift shaft is given by:

$$\text{for 8 Pa} = \frac{0.14 \times 1.61}{6} = 0.0376 \text{ m}^3/\text{s}$$

$$\text{for 50 Pa} = \frac{0.351 \times 1.61}{6} = 0.0942 \text{ m}^3/\text{s}$$

(e) *Total all the air flows out of each pressurized space and increase total by 25% in accordance with reason given in the text. This will give the air supply needed for each pressurized space.*

	8 Pa	*50 Pa*
Staircase: total air flow	0.070 m³/s	0.175 m³/s
Increase by 25%	0.0875 m³/s	0.219 m³/s
Lift lobby (each): total air flow	0.070 + 0.0376	0.175+0.0942
	= 0.1076 m³/s	0.2692 m³/s
Increase by 25%	0.134 m³/s	0.336 m³/s
Thus, total air supply needed for common system to supply six lobbies:	0.804 m³/s	2.02 m³/s

(*f*) *The air velocity through an open door should be estimated using the appropriate procedure set out in Appendix A. If the requirements are not satisfied, the air supply proposed must be increased.* (Applies to emergency level of pressurization only.)

The condition *Two doors in the same lobby open* is examined.

Case 5.A.3.2 of Appendix A applies:
(*i*) air supplied to staircase by supply duct = 0.219 m³/s
(*ii*) air flow into staircase from other lobbies

$$= \frac{0.336 \times 0.01}{0.01 + 0.03 + (\frac{1}{6} \times 0.06 \times 1.61)} \times (6 - 1)$$

$$= 0.299$$

Total air flow past staircase/lobby door	= 0.518 m³/s
Ditto corrected for resistance factor (0.6)	= 0.311 m³/s
Air velocity through this door	= 0.194 m/s

Now consider the second door, i.e. the lobby/accommodation door. Air flow past this door is given by the sum of:

(*i*) Total air flow into lobby from staircase	= 0.518 m³/s
(*ii*) Air supplied by duct to each lobby	= 0.336 m³/s
(*iii*) Air flow out of lift shaft	

$$= \frac{1}{3} \left(\frac{1.61 \times 0.06 \times 0.336}{0.01 \times 0.03 \times (\frac{1}{6} \times 0.06 \times 1.61)} \right) = 0.193 \text{ m}^3/\text{s}$$

Total air flow past door	= 1.05 m³/s
Ditto corrected for resistance factor (0.6)	= 0.628 m³/s

This gives an air velocity through door of 0.392 m/s. This is below the requirement of 0.7 m/s, therefore, extra air must be provided.

Volume air flow needed through door to give 0.7 m/s

$$= 0.7 \times 1.6 = 1.12 \text{ m}^3/\text{s}$$

Correct this for the resistance factor $= \dfrac{1.12}{0.6}$ $= 1.867 \text{ m}^3/\text{s}$

Air flow already available $= 1.05 \text{ m}^3/\text{s}$

Therefore, to achieve the required air velocity through the lobby/accommodation door, it is necessary to provide an additional 0.82 m³/s to the staircase, and in order to prevent an unacceptably high pressure developing in the stair when all doors are closed, a flap valve to relieve the pressure must be fitted.

The area of flap valve opening needs to be:

$$A = \frac{0.82}{0.827 \times 50^{1/2}} = 0.14 \text{ m}^2$$

The condition that, with only one door open, a pressure differential of 50 Pa must be developed across the closed staircase/lobby door is satisfied because of the operation of the pressure limiting flap. The open door conditions required are, therefore, satisfied.

Force exerted by a pressure of 50 Pa on a flap of area 0.14 m² is calculated as follows:

$$50 \text{ Pa} = 50 \text{ N/m}^2$$

Hence, force on flap of area 0.14 m² $= 50 \times 0.14 \text{ N} = 7 \text{ N}$.

$$1 \text{ kg wt is equal to } 9.80 \text{ N}$$

$$\text{Hence, } 7 \text{ N} = \frac{7}{9.8} \text{ kg wt} = 0.7 \text{ kg wt,}$$

which is the force which must be designed into the pressure relief flap to hold it in the closed position.

(g) *The air supply calculated in the design [(e) and (f) above] must be provided at the duct terminal (or terminals) in each pressurized space. The positions of the duct terminals should be discussed with the architect and the appropriate authorities.*

An air supply grille will be required in each lobby and at least two, but preferably three, in the staircase. The ductwork must be so positioned in the building that fire dampers are not necessary.

(h) *The fan capacity and duct sizes must be decided by a competent engineer after due consideration of any additional requirements. The position of the intake grilles must be agreed with the architect and any special protection to the installation specified.*

Emergency condition (50 Pa):
The supply required is   1.04 m³/s to the staircase
                  and   2.02 m³/s for six lobbies
making a total supply of   3.06 m³/s

Normal condition (8 Pa):
The supply required is   0.09 m³/s to the staircase
                  and   0.81 m³/s for six lobbies
making a total supply of   0.9 m³/s

At this stage the decision must be taken as to whether staircase and lobbies are to be supplied from one or two fans, and any requirement concerning standby equipment and alternative electrical supply discussed. If the two-stage system uses separate equipment for each stage, the air intake position need not be the same for both stages, but a position at ground level is preferred for the emergency stage.

(*i*) *The escape of the pressurizing air from the building should be considered and the appropriate method of venting specified.*
   In this example the air release venting from the accommodation must be sufficient to cater for the air passing an open door at 0.7 m/s, i.e.

$$Q_N = 0.7 + 1.6 = 1.12 \ \text{m}^3/\text{s}$$

The vent size and distribution will depend on which of the four methods are chosen, but due note must be taken of the measures which refer to the adverse wind condition.
   (*j*) *The operation of the system must be considered and the position of smoke detectors (if required) specified.*
   This point must be discussed with the Building Control and/or Fire Authorities at an early stage. A smoke detector on the accommodation side of the lobby door on each floor is probably a minimum requirement, unless manual alarm is proposed.
   (*k*) *A note of the leakage areas assumed must be given to the architect reminding him that these areas must be achieved in the finished building.*
   Any important divergence from the assumed leakage areas will become apparent at the acceptance test stage but preferably they should be verified before this time.
   (*l*) *A measurement procedure should be specified so that satisfactory operation of the installation in the completed building can be established.*
   The new Code of Practice lays down the measurements required and describes the measurement techniques to be used. These methods should be followed in the tests on the completed installation.

Index

174